walkermaths 1.6

GEOMETRY

NCEA Level 1 External

Charlotte Walker and Victoria Walker

Walker Maths 1.6 Geometry
1st Edition
Charlotte Walker
Victoria Walker

Editor: Eva Chan
Designer: Cheryl Smith, Macarn Design
Production controller: Siew Han Ong

Any URLs contained in this publication were checked for currency during the production process. Note, however, that the publisher cannot vouch for the ongoing currency of URLs.

Acknowledgements
Cover photo courtesy of Shutterstock.

We wish to thank the Boards of Trustees of Darfield and Riccarton High Schools for allowing us to use materials and ideas developed while teaching. Our thanks also go to all past and present colleagues who have generously shared their expertise and ideas.

For product information and technology assistance,
in Australia call **1300 790 853**;
in New Zealand call **0800 449 725**

For permission to use material from this text or product, please email **aust.permissions@cengage.com**

National Library of New Zealand Cataloguing-in-Publication Data
A catalogue record for this book is available from the National Library of New Zealand.

ISBN 978 0 17 037039 4

Cengage Learning Australia
Level 7, 80 Dorcas Street
South Melbourne, Victoria Australia 3205

Cengage Learning New Zealand
Unit 4B Rosedale Office Park
331 Rosedale Road, Albany, North Shore 0632, NZ

For learning solutions, visit **cengage.co.nz**

Printed in China by 1010 Printing International Limited
12 13 14 25 24 23

CONTENTS

ISBN: 9780170370394

Glossary

Make your own glossary of key terms:

Term	Definition	Picture/Example
The Theorem of Pythagoras		
Hypotenuse		
Opposite		
Adjacent		
Sine (sin)		
Cosine (cos)		
Tangent (tan)		
Equilateral triangle		
Isosceles triangle		
Scalene triangle		

ISBN: 9780170370394

Sector		
Segment		
Arc		
Similar shape		
Congruent		
Tessellate		
Vertical		
Horizontal		
Intersect		
Bisect		
Adjacent		
Perpendicular		
Plane		

ISBN: 9780170370394

The Theorem of Pythagoras

- The Theorem of Pythagoras applies to **right-angled** triangles only.
- The longest side is the **hypotenuse**.
- The hypotenuse is always **opposite** the right angle.
- The theorem is used for finding **lengths** of sides.

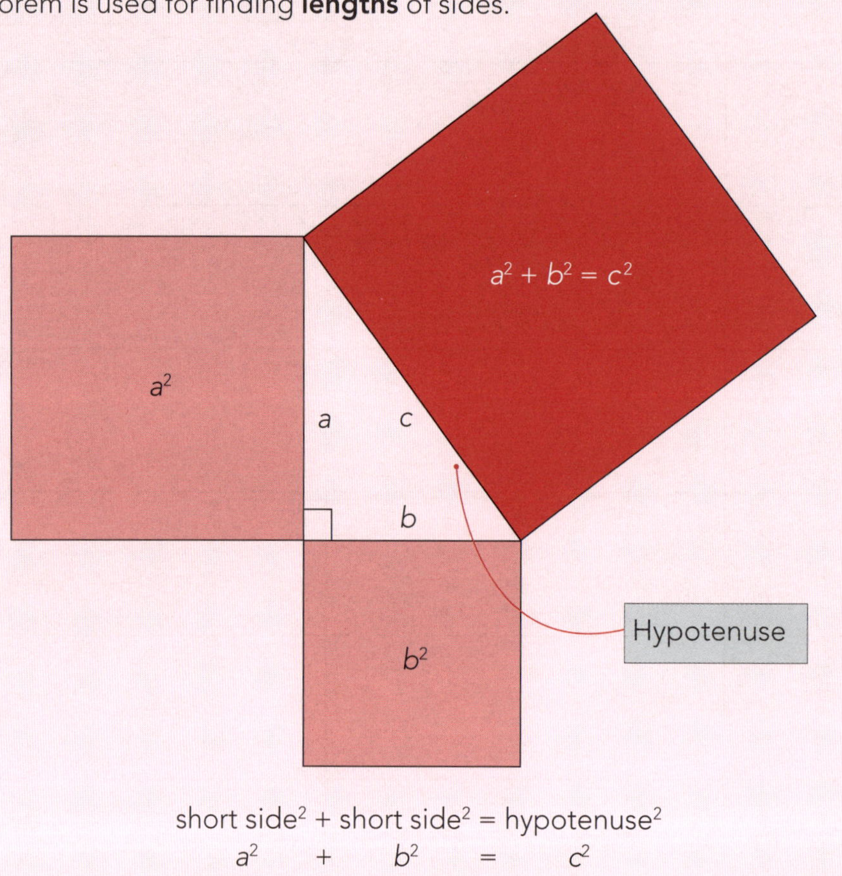

short side^2 + short side^2 = hypotenuse^2

$$a^2 \quad + \quad b^2 \quad = \quad c^2$$

Finding the length of the hypotenuse

Example: Calculate the length of *c*.

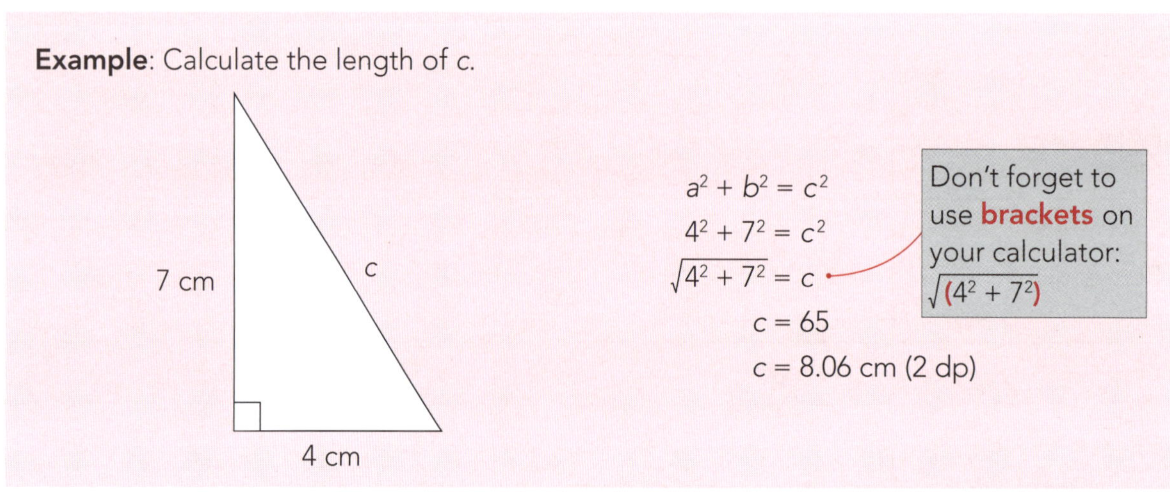

$$a^2 + b^2 = c^2$$
$$4^2 + 7^2 = c^2$$
$$\sqrt{4^2 + 7^2} = c$$
$$c = 65$$
$$c = 8.06 \text{ cm (2 dp)}$$

ISBN: 9780170370394

Calculate the length of the unknown side of each triangle.

1

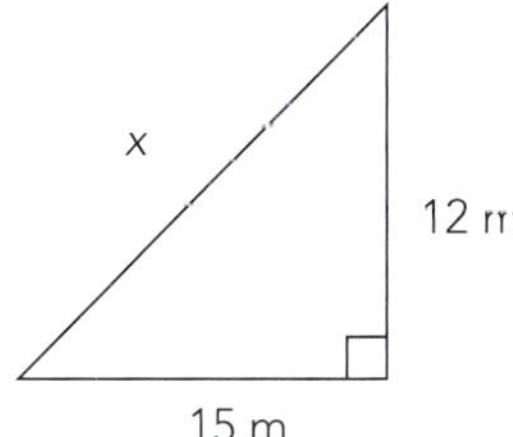

2

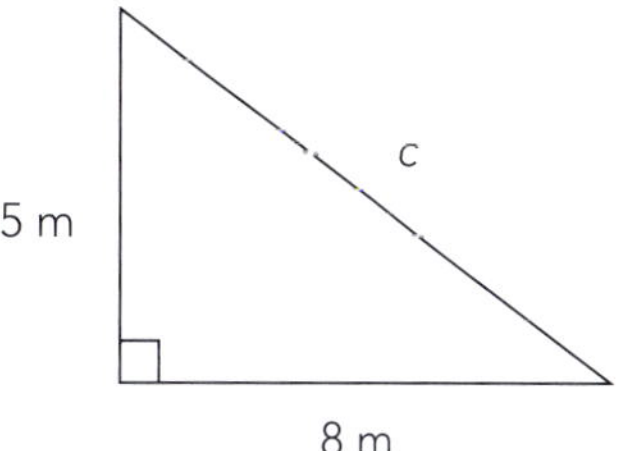

3

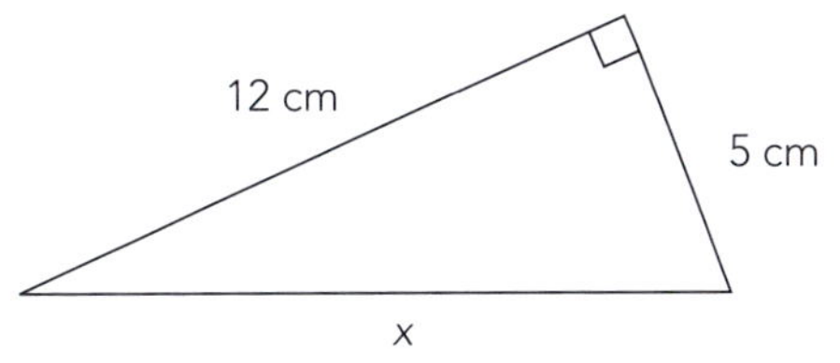

4

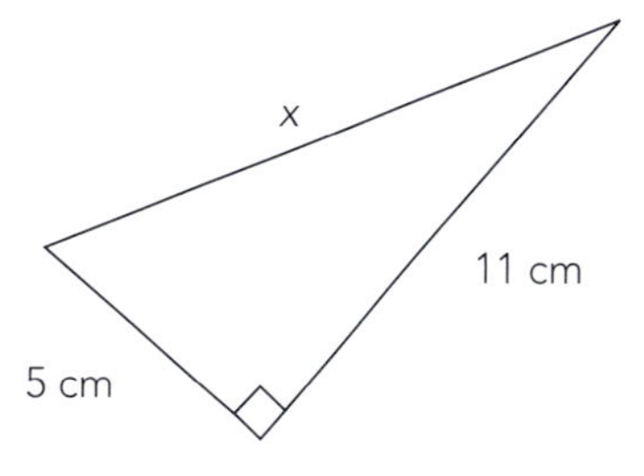

5

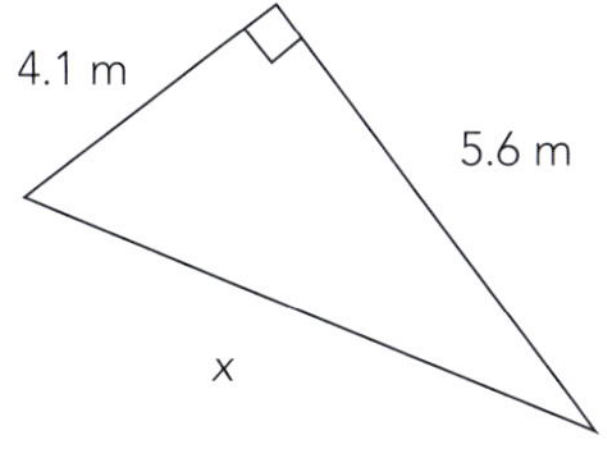

6

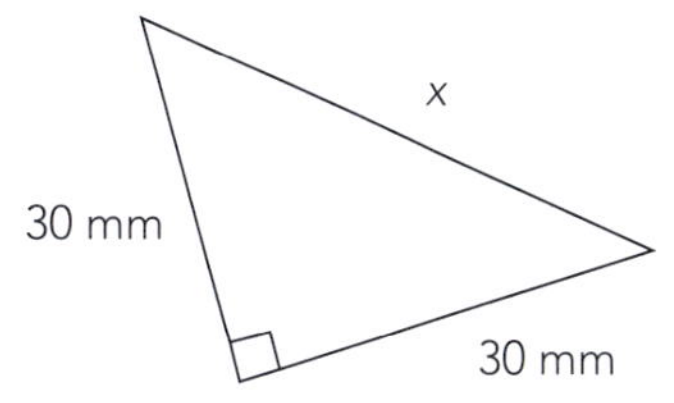

7

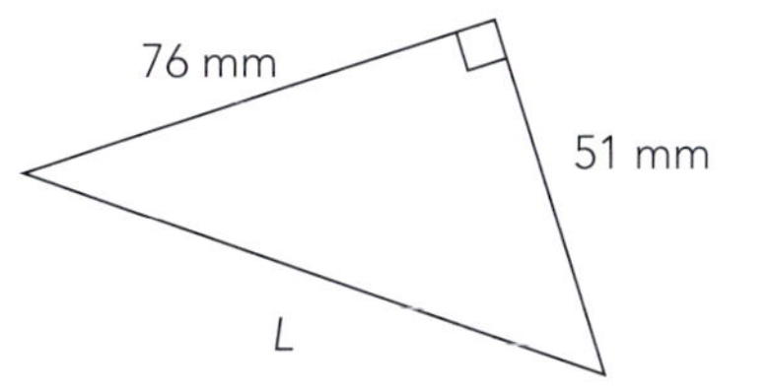

8

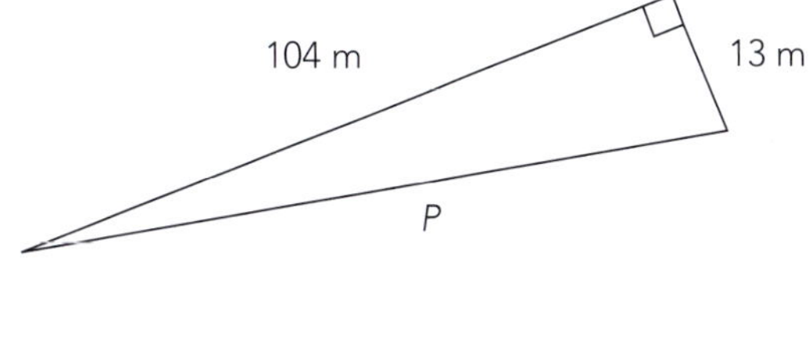

ISBN: 9780170370394

Finding the lengths of short sides

Example: Calculate the length of y.

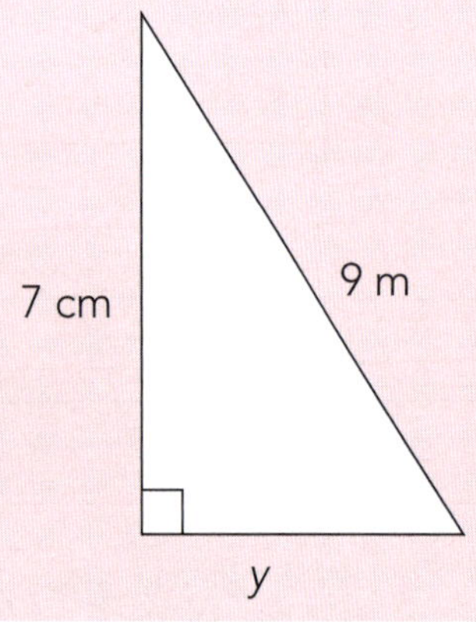

$a^2 + b^2 = c^2$

$y^2 + 7^2 = 9^2$

$y^2 = 9^2 - 7^2$

$y = \sqrt{9^2 - 7^2}$

$y = 5.66$ m (2 dp)

Don't forget to use **brackets** on your calculator: $\sqrt{(9^2 - 7^2)}$

Calculate the length of the unknown side of each triangle.

1

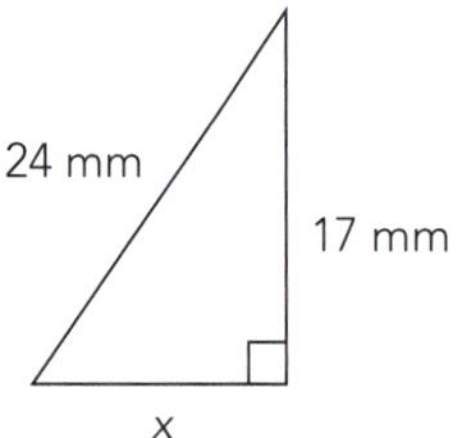

2

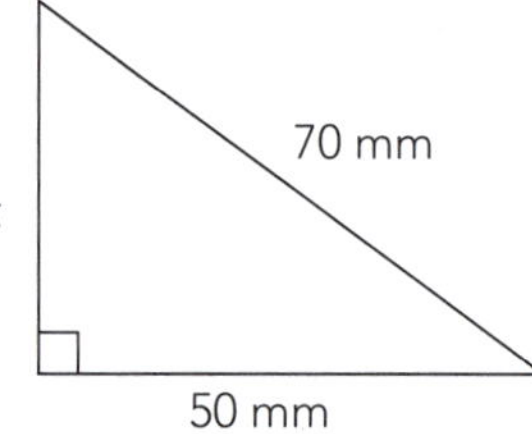

3

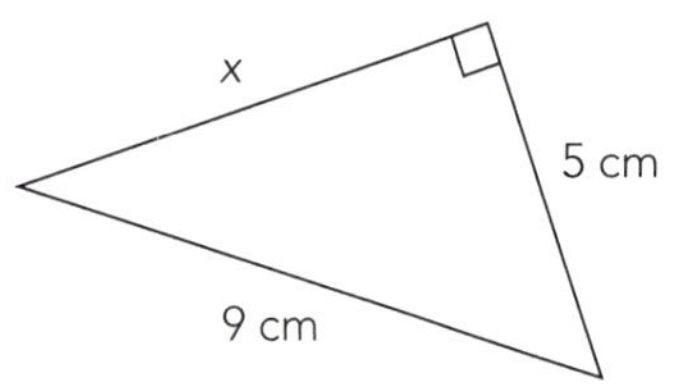

4

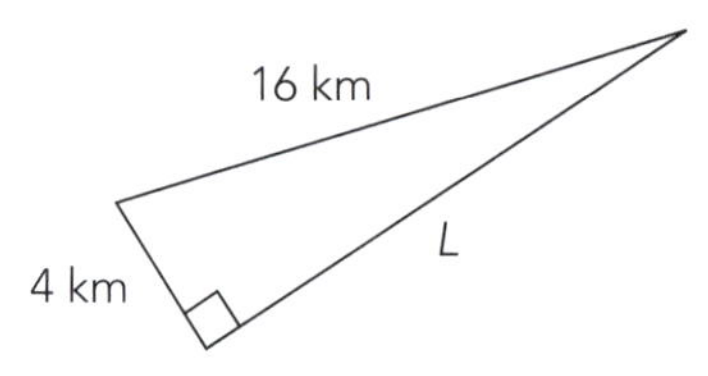

5

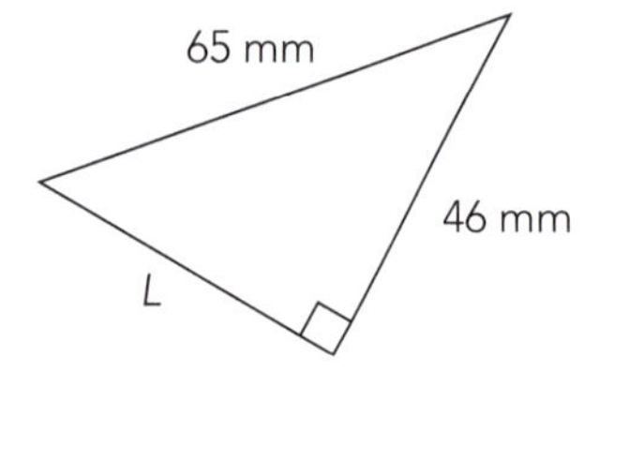

6

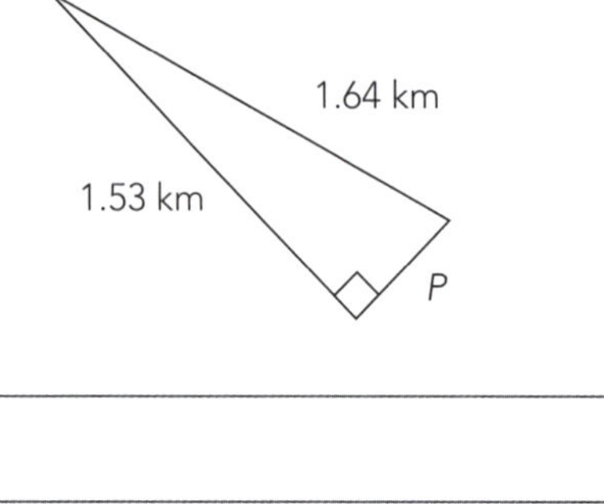

 ISBN: 9780170370394

Mixing it up

Calculate the length of the unknown side of each triangle.

1

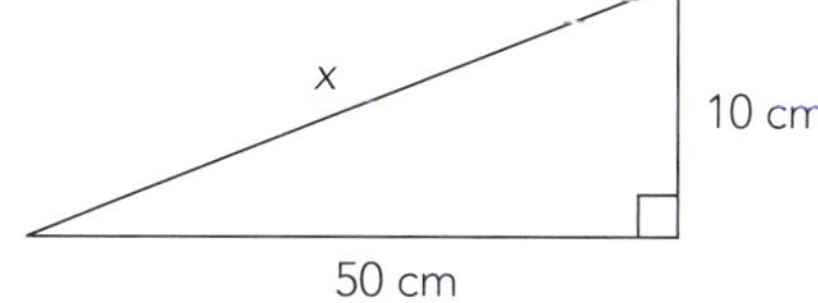

2

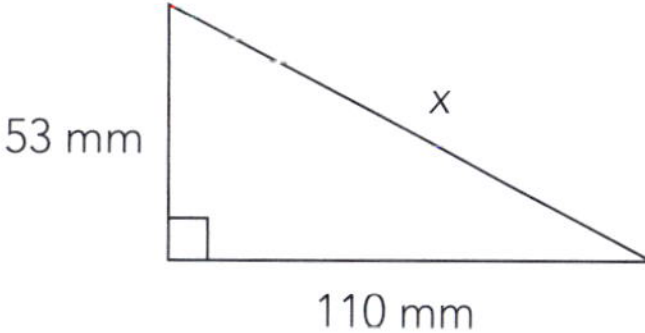

3

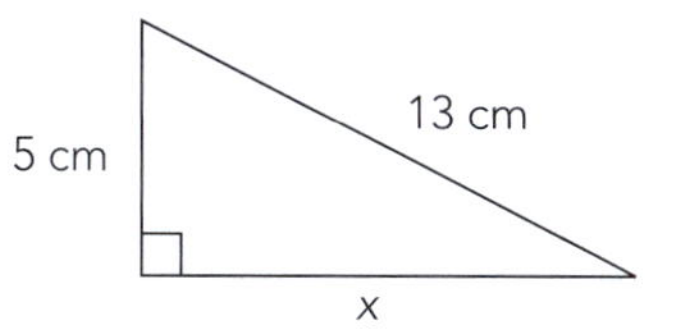

4

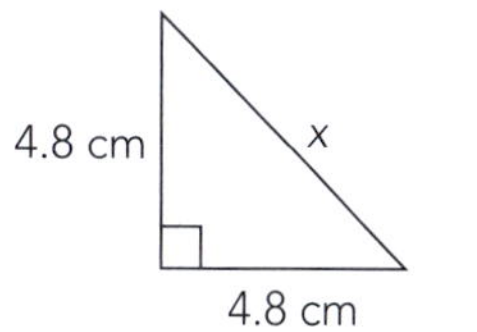

5

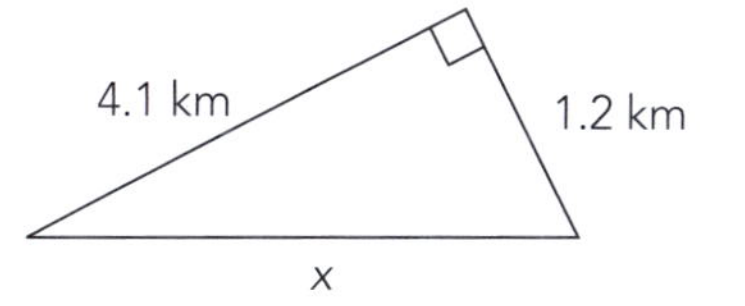

6

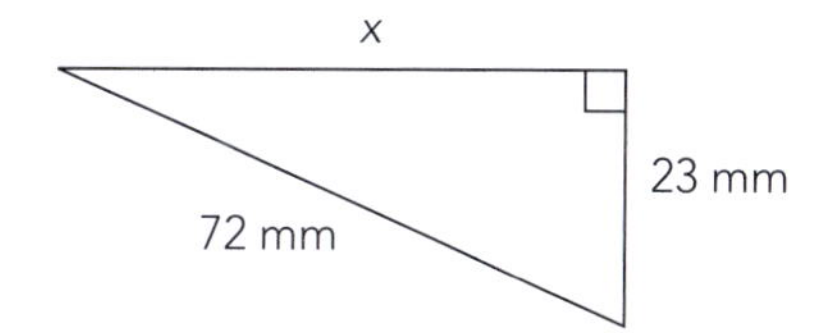

7

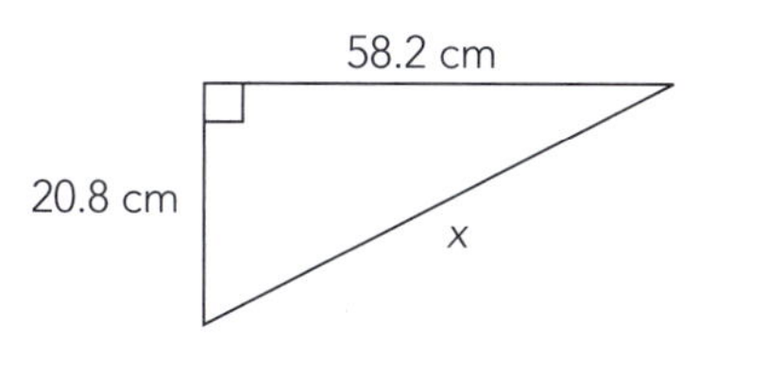

8

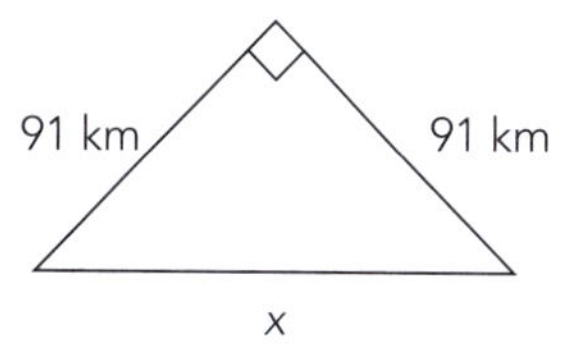

9

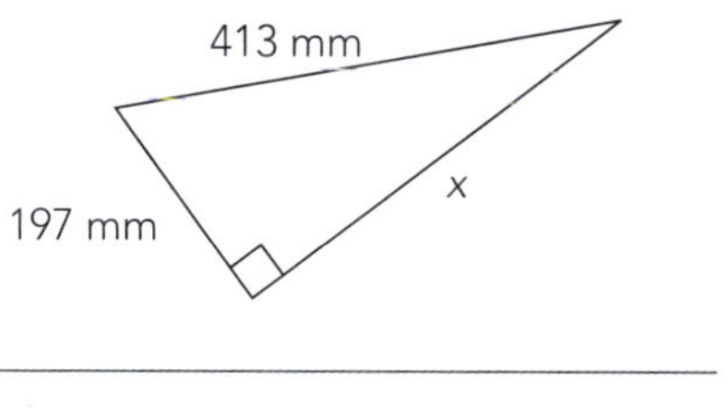

10

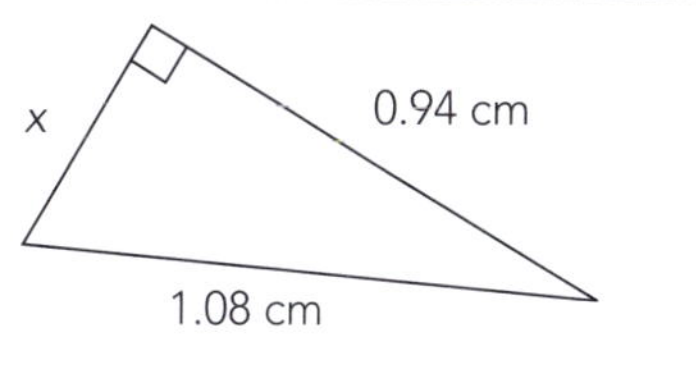

ISBN: 9780170370394

Applications

1 ABCD is a square.
Calculate the length of AC.
How long is AE?

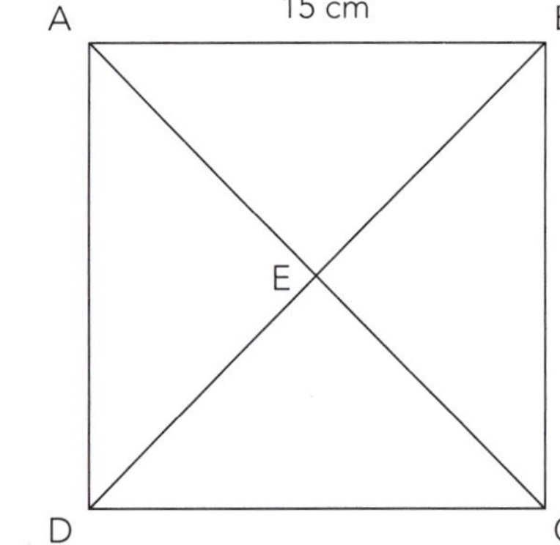

2 ABCD is a rectangle. AC is 95 mm.
Calculate the length of AB.

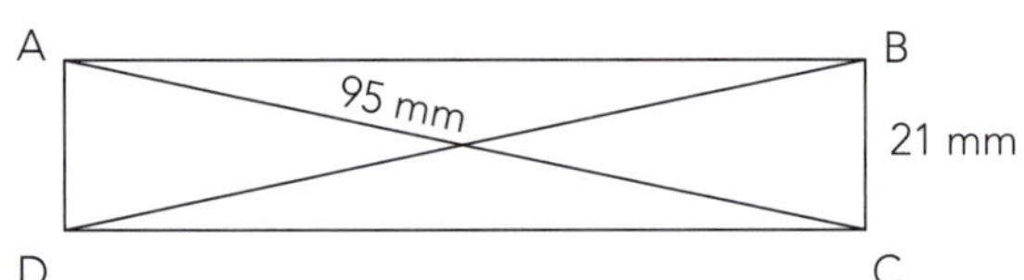

3 ABCD is a trapezium.
Calculate the length of AB.

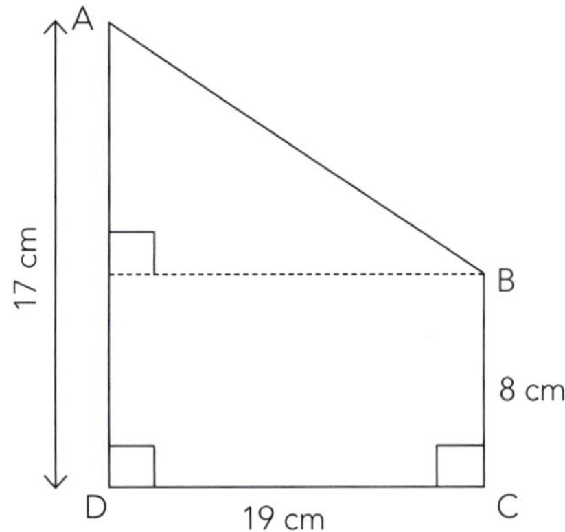

4 ABCD is a trapezium.
Calculate the length of BC.

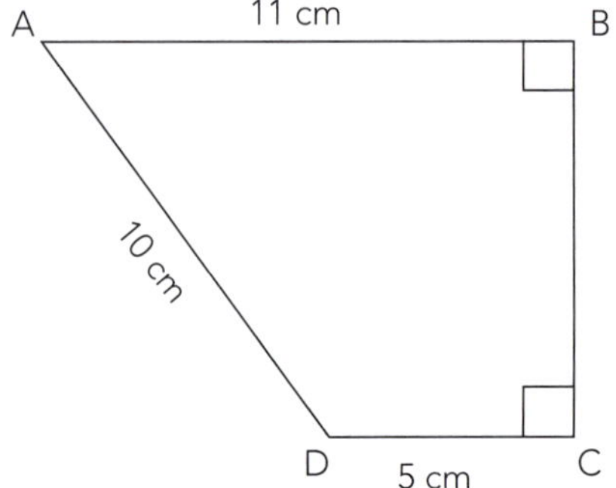

ISBN: 9780170370394

5 Both ABC and ADB are isosceles triangles.
Calculate the perimeter of triangle ABC.

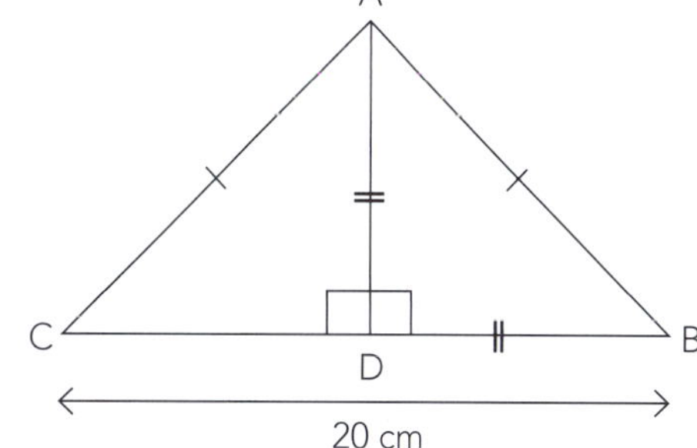

____

____

____

6 A ladder is leaning against a wall.
The base of the ladder is 1.6 m from the wall, and the ladder reaches 3.6 m up the wall.
Calculate the length of the ladder.

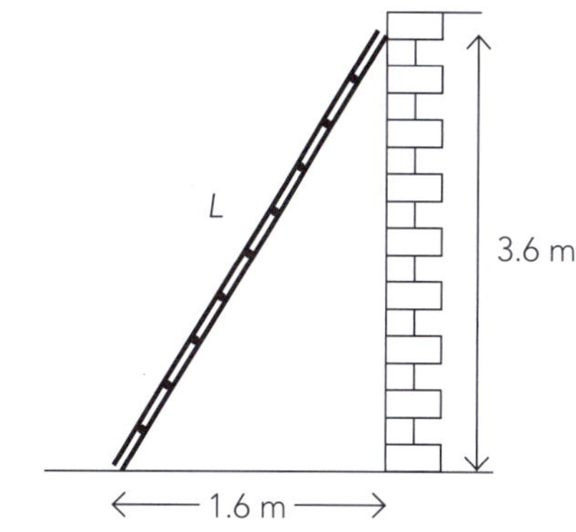

____

____

____

7 A 4 m long ladder is leaning against a wall.
If the base of the ladder is 1.4 m from the wall, calculate how far up the wall the ladder reaches.

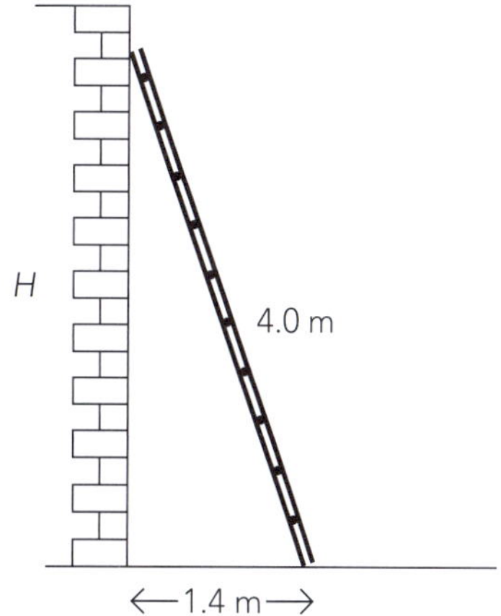

____

____

____

8 Calculate the length of AB.

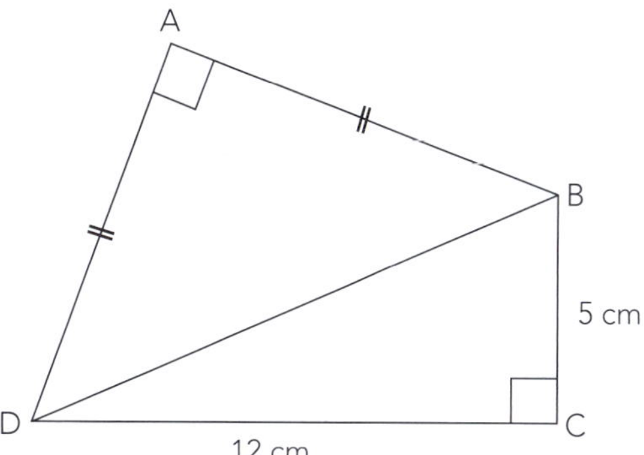

____

____

____

ISBN: 9780170370394

Trigonometry

- Trigonometry can also be used for calculations involving **right-angled triangles**.
- However, unlike calculations using the Theorem of Pythagoras, an **angle** must be involved.

Before you begin a calculation involving trigonometry, **label** the sides of the triangle:

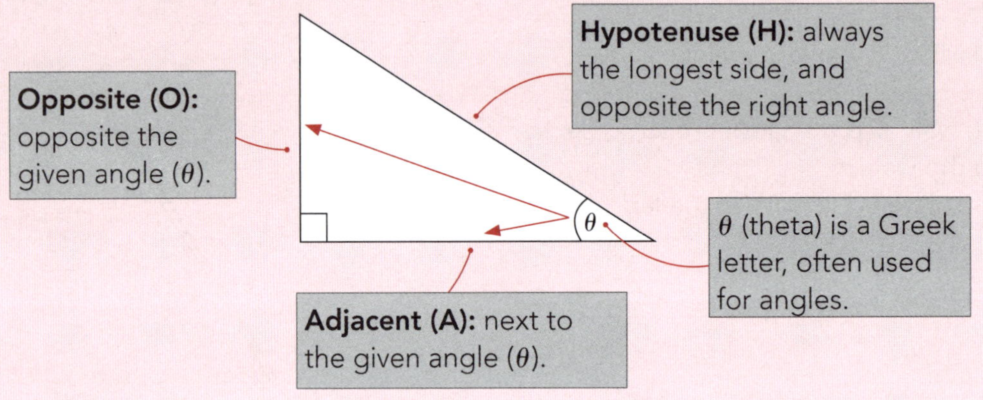

Label the following triangles with H, O and A.

1

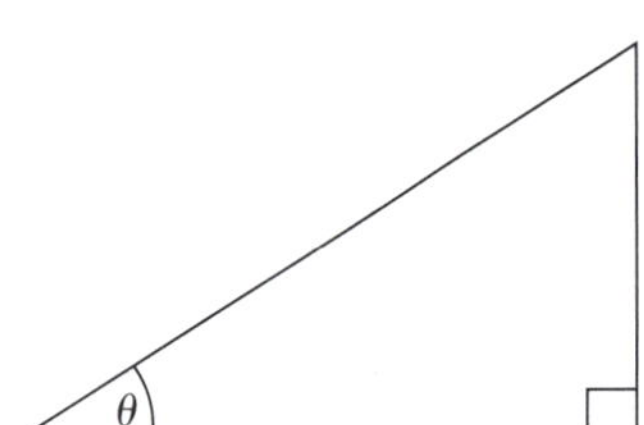

2

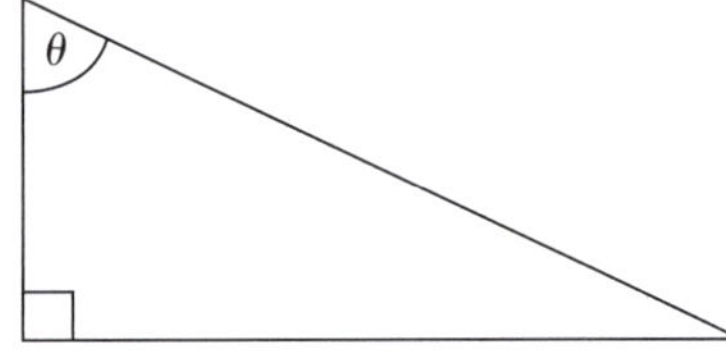

3

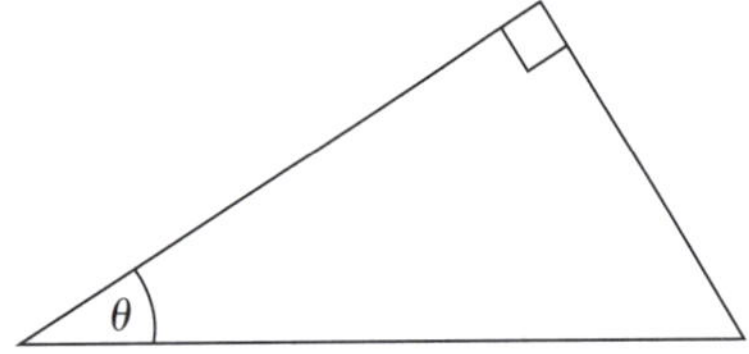

4

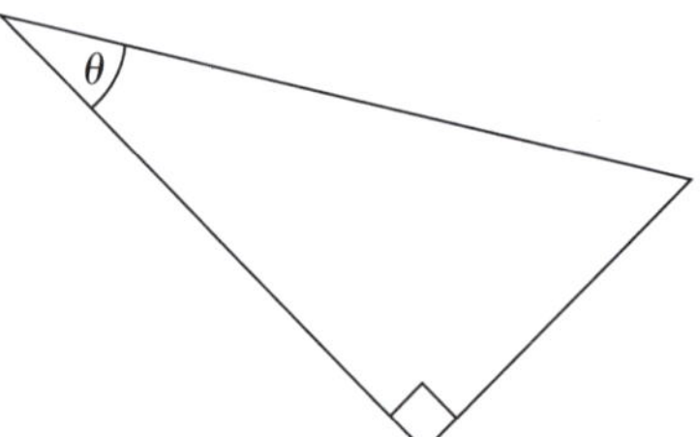

5

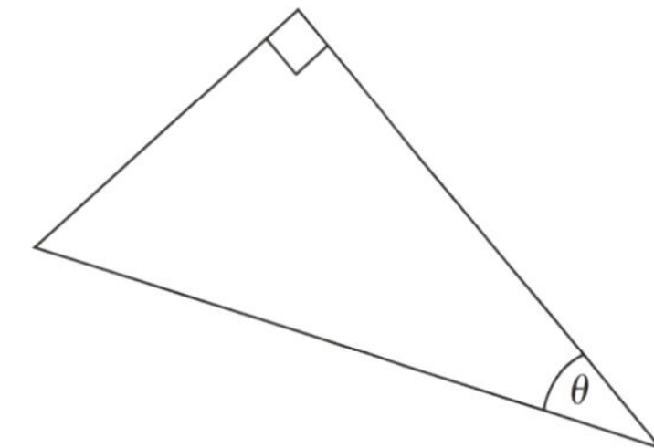

6

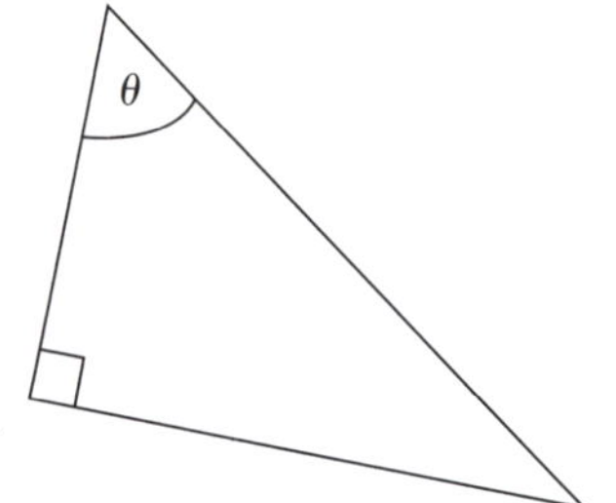

ISBN: 9780170370394

SOH CAH TOA

You will need to know about three trigonometrical functions: **sin** θ (sine)
cos θ (cosine)
tan θ (tangent).

These are the rules in trigonometry:

$$\sin\theta = \frac{O}{H} \qquad \cos\theta = \frac{A}{H} \qquad \tan\theta = \frac{O}{A}$$

These are usually remembered as the 'word' **SOH CAH TOA**.

Organising **SOH CAH TOA** into triangles can help you work out how to use it:

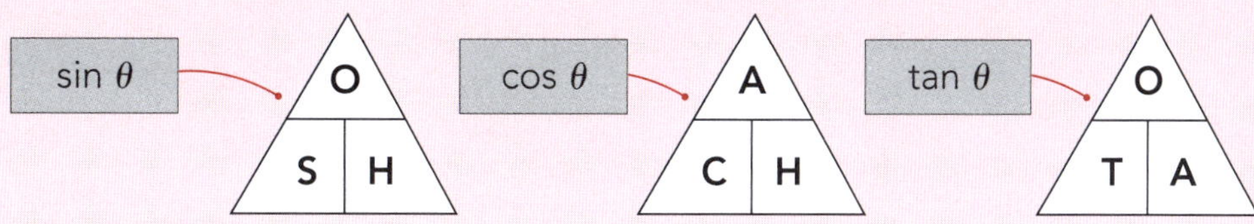

Finding short sides using sine and cosine

Example one: Calculate the length marked x.

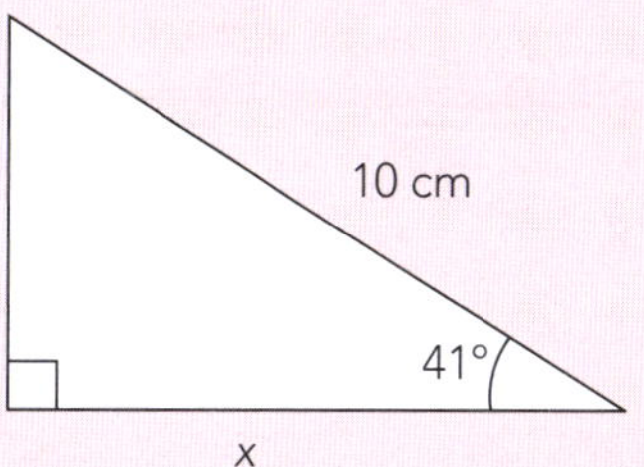

Step 1: **Label** the sides that are involved in the problem with 'A', 'O' and 'H'.

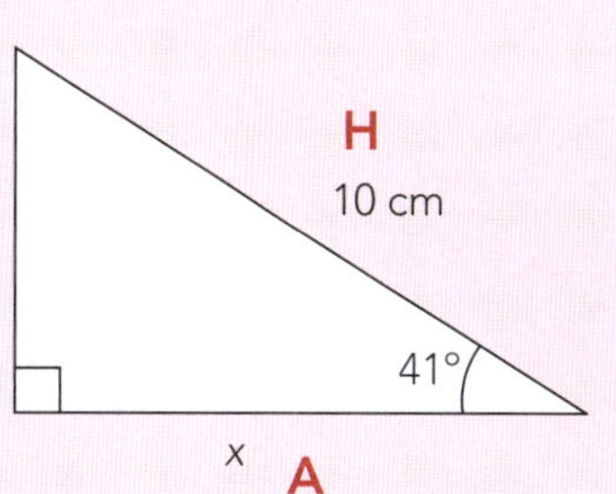

Step 2: Decide **which relationship** involves these sides. In this case we are concerned only with A and H so we must use:

$$\cos\theta = \frac{A}{H}$$

Step 3: **Substitute** the values from the triangle.

$$\cos 41^\circ = \frac{x}{10}$$

$$x = 10\cos 41^\circ$$

$$= 7.547 \text{ cm}$$

Check that your calculator is set to **degrees**.

Step 4: **Think about your answer — does it seem reasonable?**
In this case, the hypotenuse is 10 cm and x must be shorter, so an answer of 7.547 cm is reasonable.

ISBN: 9780170370394

Example two: Calculate the length marked x.

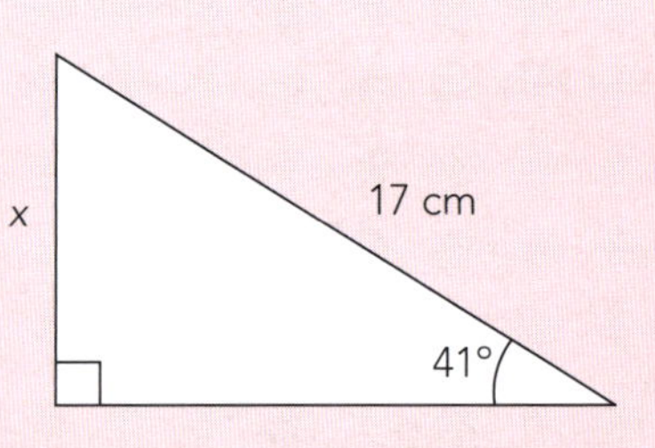

Step 1: **Label** the sides that are involved in the problem with 'A', 'O' and 'H'.

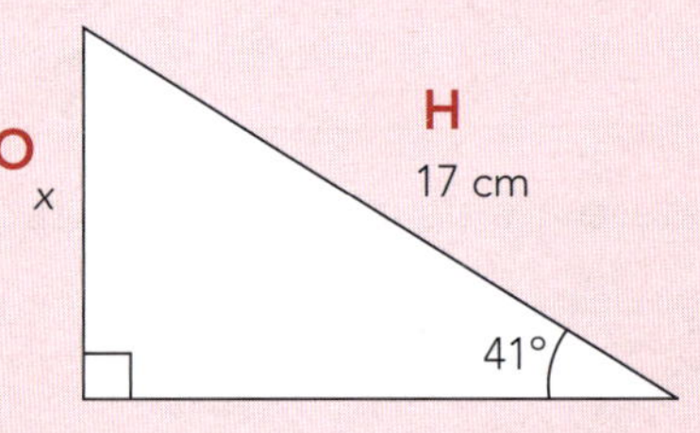

Step 2: Decide **which relationship** involves these sides. In this case we are concerned only with O and H so we must use:

$$\sin \theta = \frac{O}{H}$$

Step 3: **Substitute** the values from the triangle.

$$\sin 41° = \frac{x}{17}$$

$$x = 17\sin 41°$$

$$= 11.15 \text{ cm}$$

Step 4: **Think about your answer — does it seem reasonable?**
In this case, the hypotenuse is 17 cm and x must be shorter, so an answer of 11.15 cm is reasonable.

Calculate the length of the unknown side of each triangle.

1

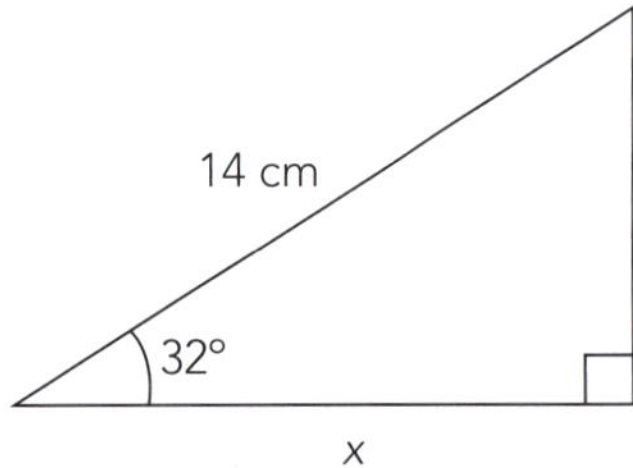

2

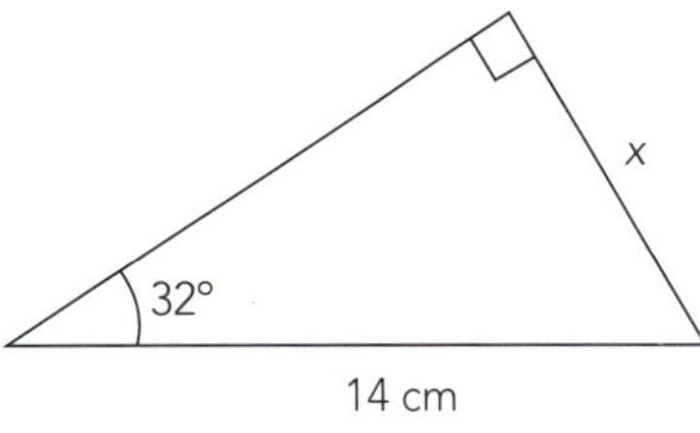

3

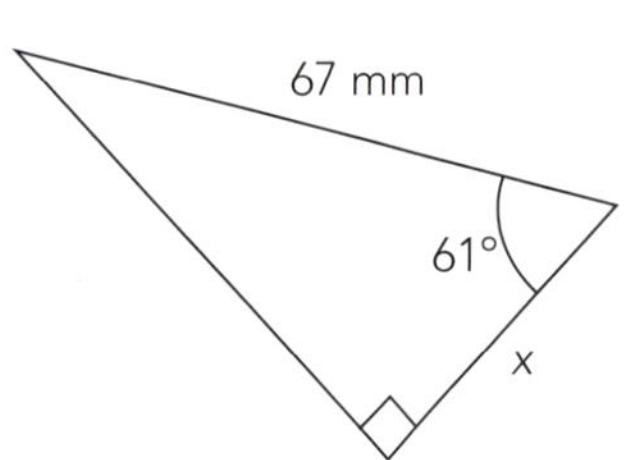

ISBN: 9780170370394

4

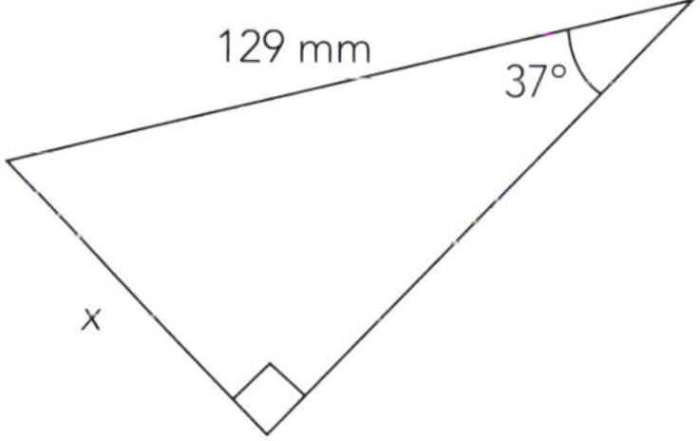

5

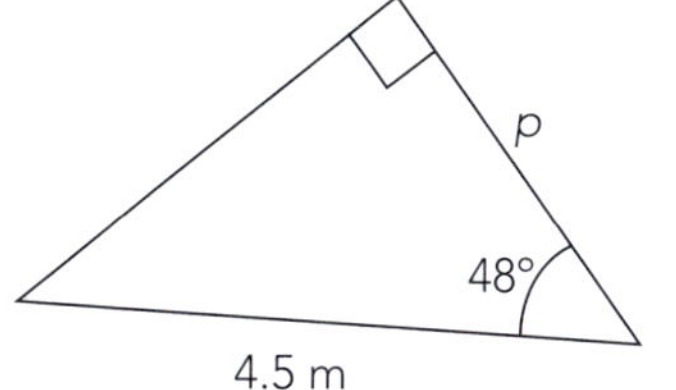

6

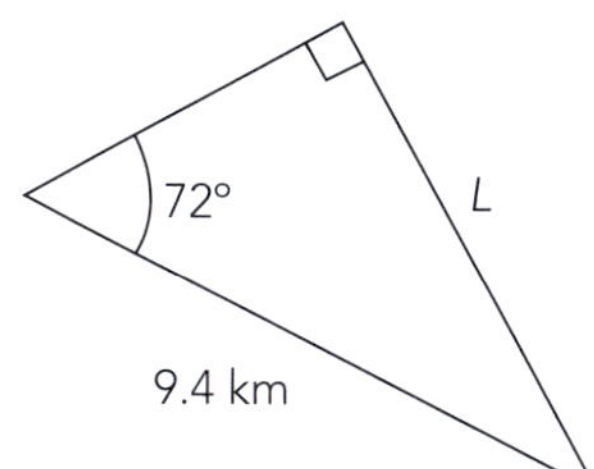

7

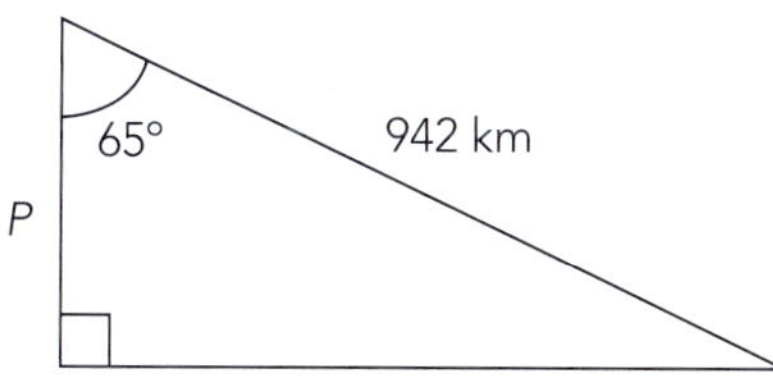

8

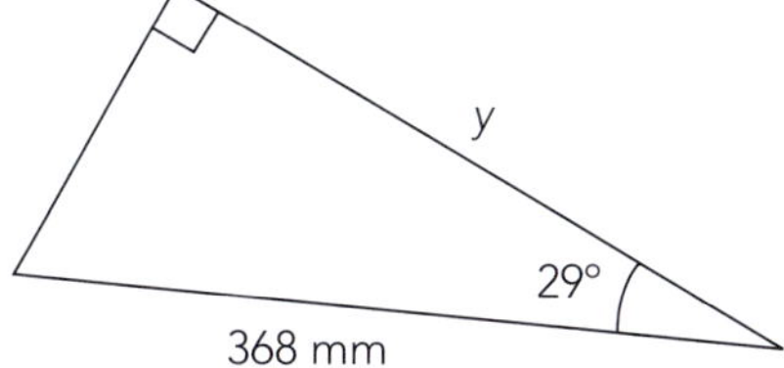

9

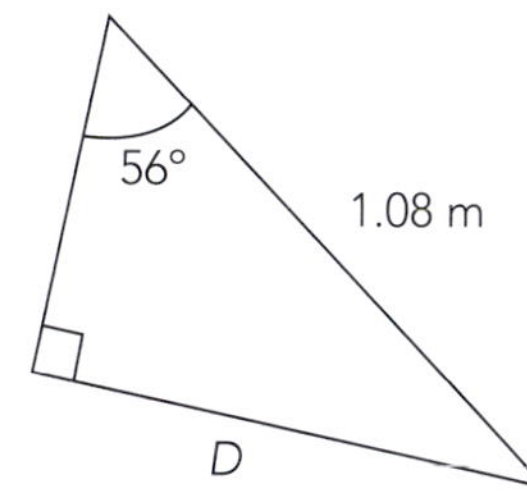

10

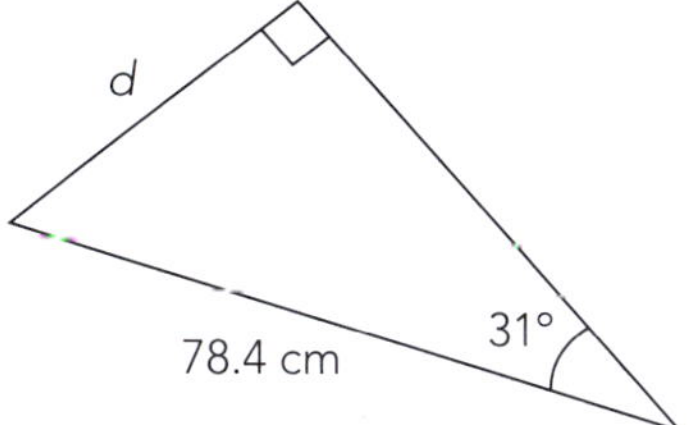

ISBN: 9780170370394

Finding long sides (hypotenuse) using sine and cosine

This is the same process, but the algebra is a little different.

Example one: Calculate the length marked x.

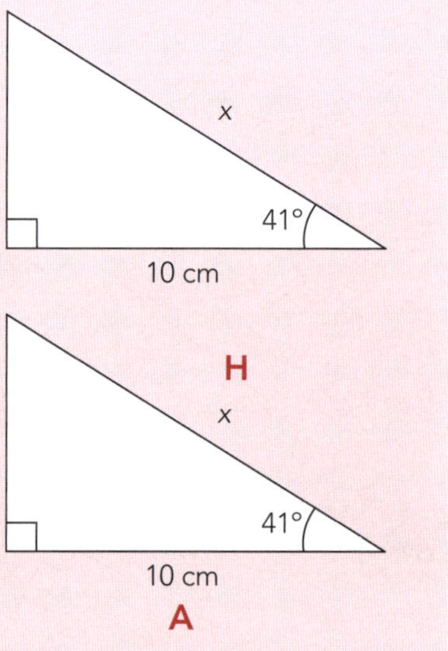

Step 1: **Label** the sides that are involved in the problem with 'A', 'O' and 'H'.

Step 2: Decide **which relationship** involves these sides. In this case we are concerned only with A and H so we must use:

$$\cos\theta = \frac{A}{H}$$

Step 3: **Substitute** the values from the triangle.

$$\cos 41° = \frac{10}{x}$$

$$x \cos 41° = 10$$

$$x = \frac{10}{\cos 41°}$$

$$x = 13.25 \text{ cm}$$

Step 4: **Think about your answer — does it seem reasonable?**
In this case, the adjacent side is 10 cm and x must be longer, so an answer of 13.25 cm is reasonable.

Example two: Calculate the length marked x.

17 cm x 41°

Step 1: **Label** the sides that are involved in the problem with 'A', 'O' and 'H'.

Step 2: Decide **which relationship** involves these sides. In this case we are concerned only with O and H so we must use:

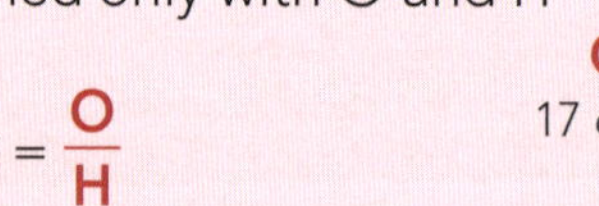

$$\sin\theta = \frac{O}{H}$$

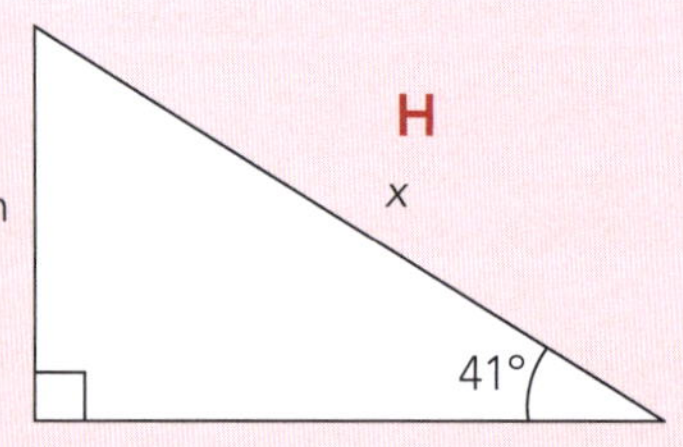

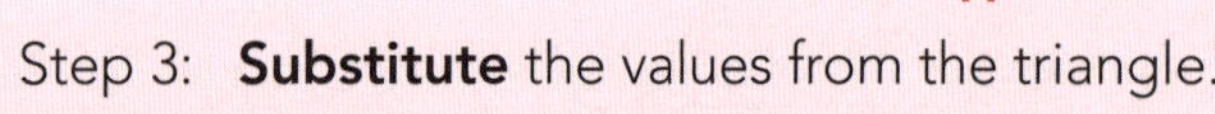

Step 3: **Substitute** the values from the triangle.

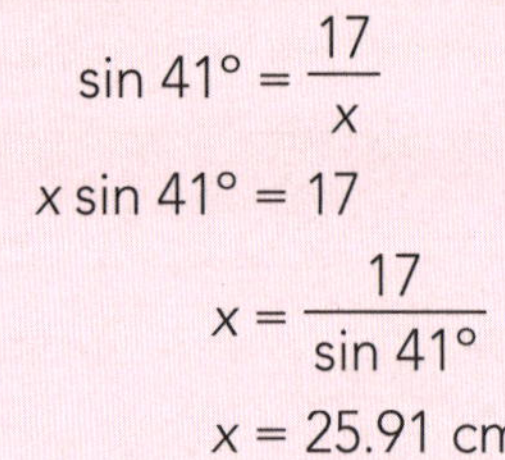

$$\sin 41° = \frac{17}{x}$$

$$x \sin 41° = 17$$

$$x = \frac{17}{\sin 41°}$$

$$x = 25.91 \text{ cm}$$

Step 4: **Think about your answer — does it seem reasonable?**
In this case, the opposite side is 17 cm and x must be longer, so an answer of 25.91 cm is reasonable.

ISBN: 9780170370394

Calculate the length of the unknown side of each triangle.

1

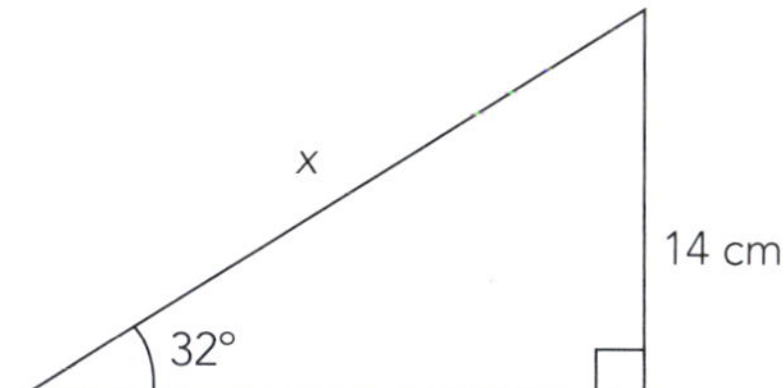

2

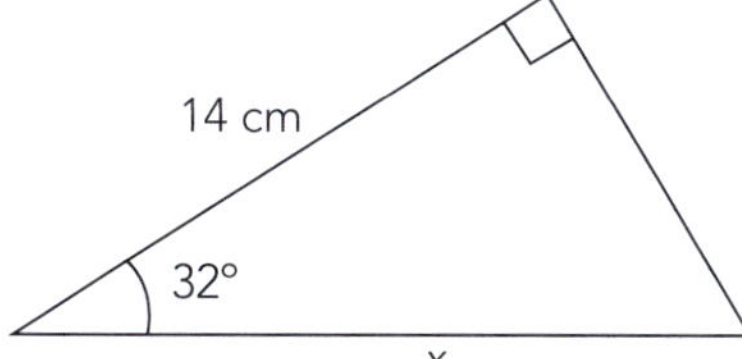

3

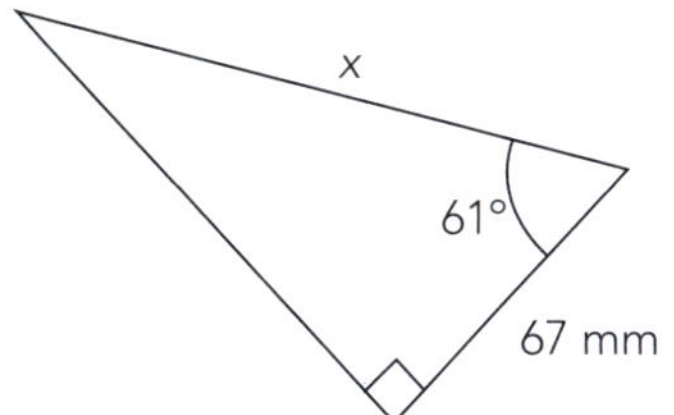

4

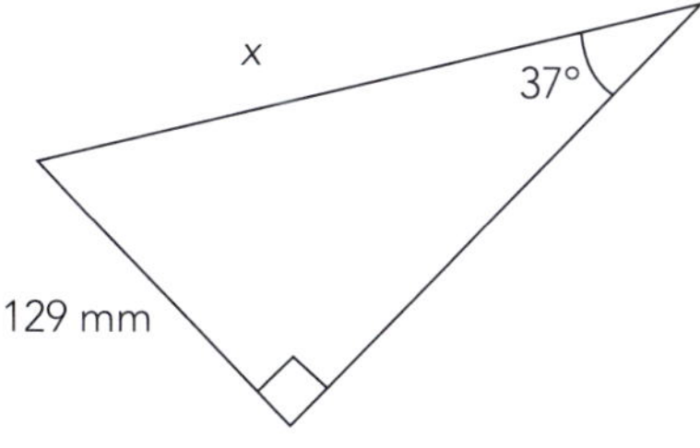

5

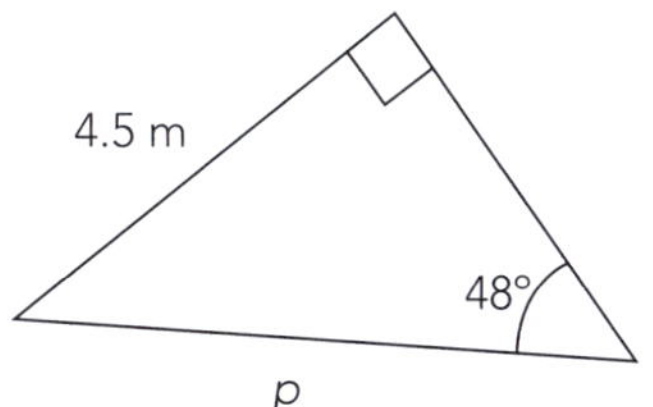

6

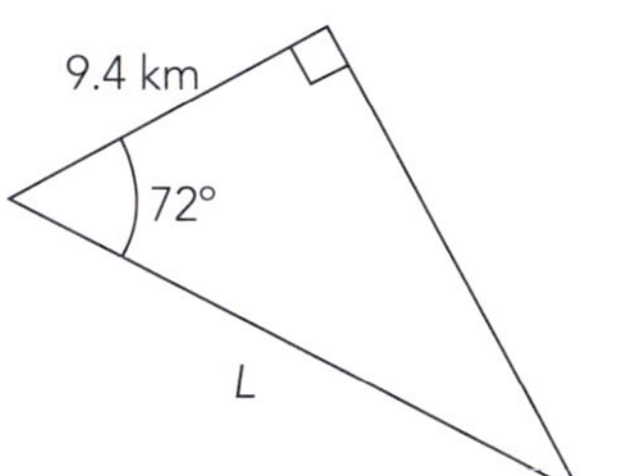

7

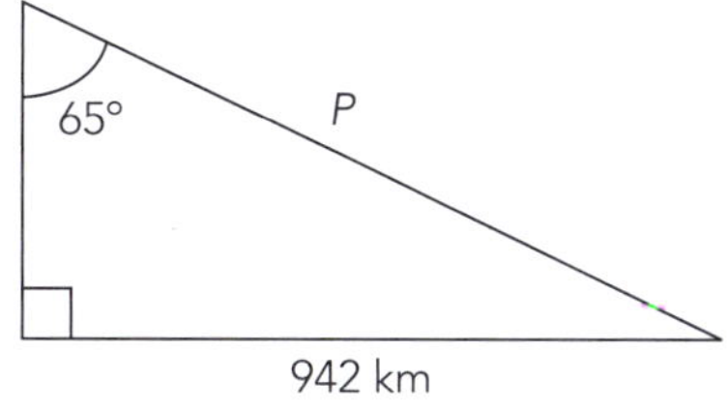

ISBN: 9780170370394

Finding sides using tangents

This involves using either of the algebraic methods used earlier.

Example one: Calculate the length marked x.

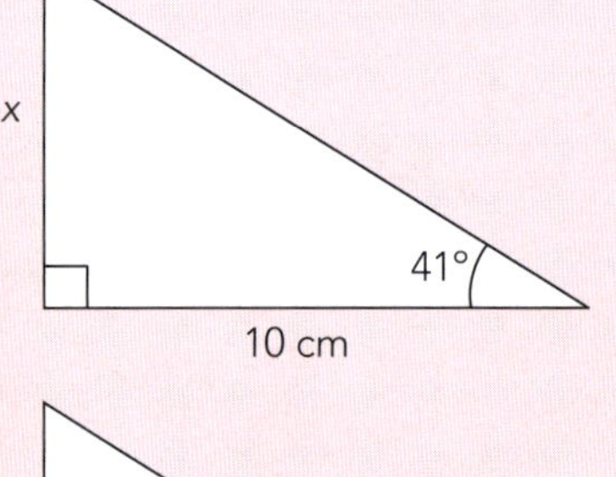

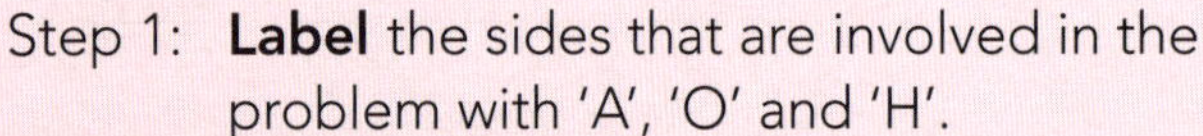

Step 1: **Label** the sides that are involved in the problem with 'A', 'O' and 'H'.

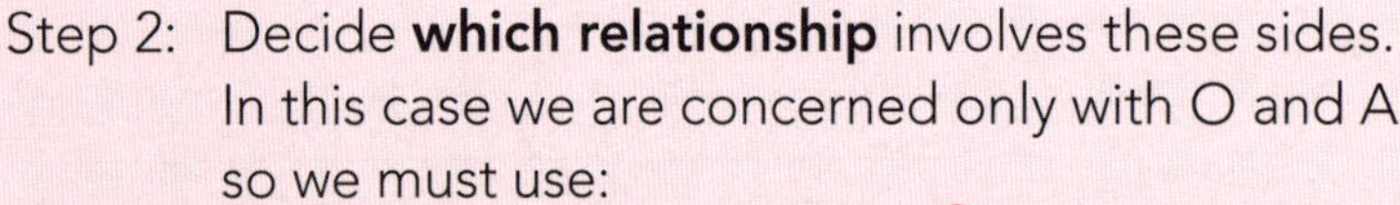

Step 2: Decide **which relationship** involves these sides. In this case we are concerned only with O and A so we must use:

$$\tan\theta = \frac{O}{A}$$

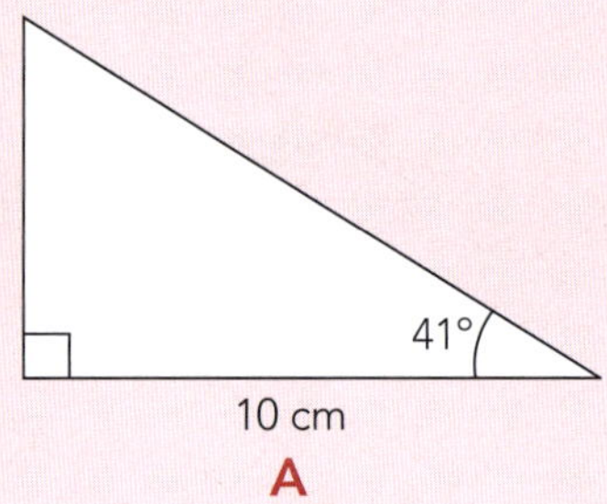

Step 3: **Substitute** the values from the triangle.

$$\tan 41° = \frac{x}{10}$$

$$x = 10 \tan 41°$$

$$x = 8.693 \text{ cm}$$

Step 4: **Think about your answer — does it seem reasonable?**
In this case, x must be less than 10 cm because x is opposite the smallest angle (41°), so an answer of 8.693 cm is reasonable.

Example two: Calculate the length marked x.

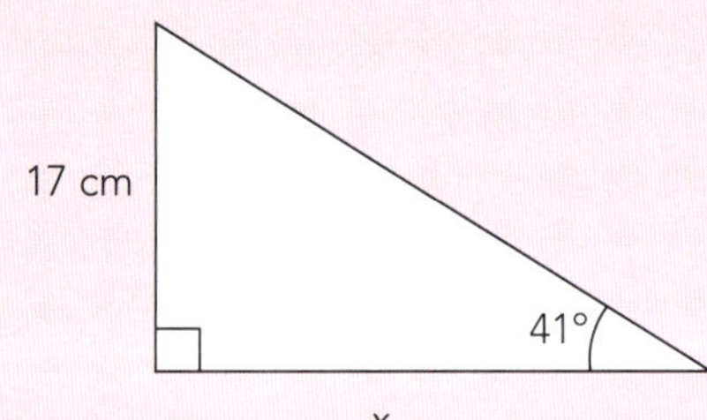

Step 1: **Label** the sides that are involved in the problem with 'A', 'O' and 'H'.

Step 2: Decide **which relationship** involves these sides. In this case we are concerned only with O and A so we must use:

$$\tan\theta = \frac{O}{A}$$

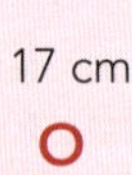

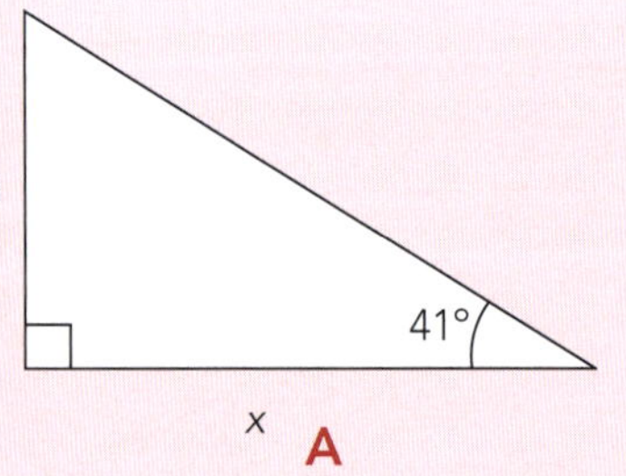

Step 3: **Substitute** the values from the triangle.

$$\tan 41° = \frac{17}{x}$$

$$x \tan 41° = 17$$

$$x = \frac{17}{\tan 41°}$$

$$x = 19.56 \text{ cm}$$

Step 4: **Think about your answer — does it seem reasonable?**
In this case, x must be more than 17 cm because the 17 cm is opposite the smallest angle (41°), so an answer of 19.56 cm is reasonable.

ISBN: 9780170370394

Calculate the length of the unknown side of each triangle.

1

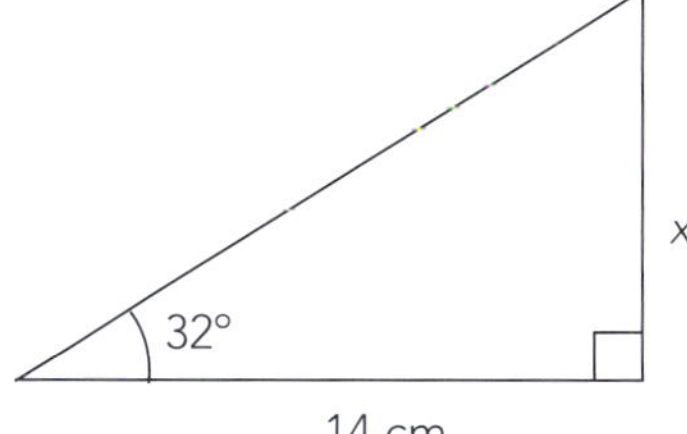

2

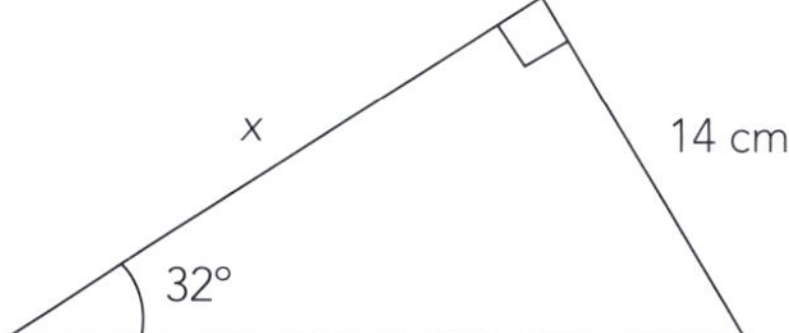

3

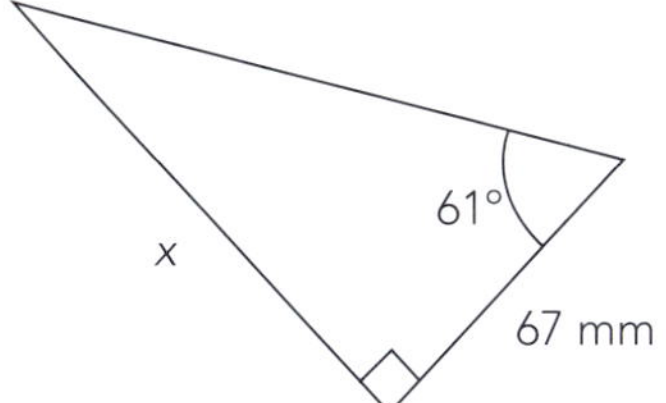

4

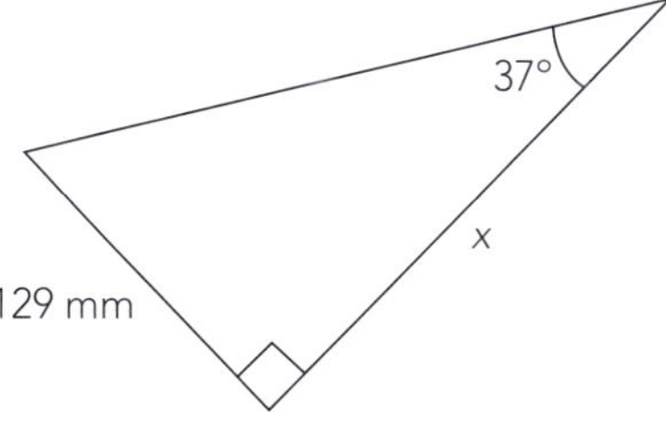

5

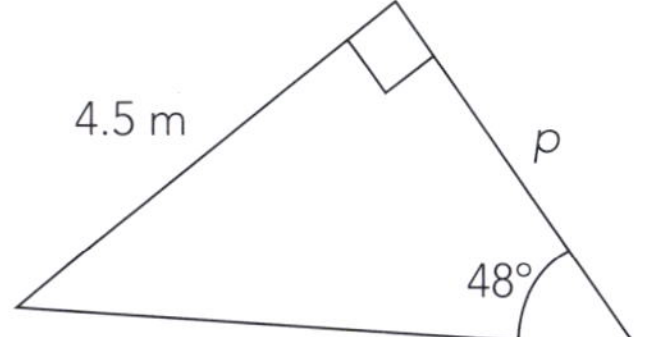

6

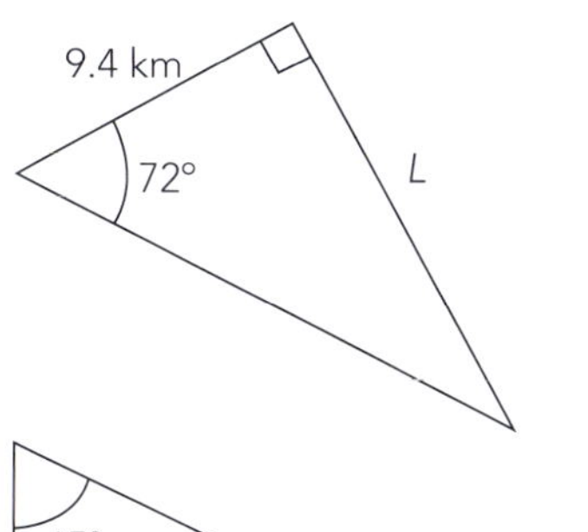

7

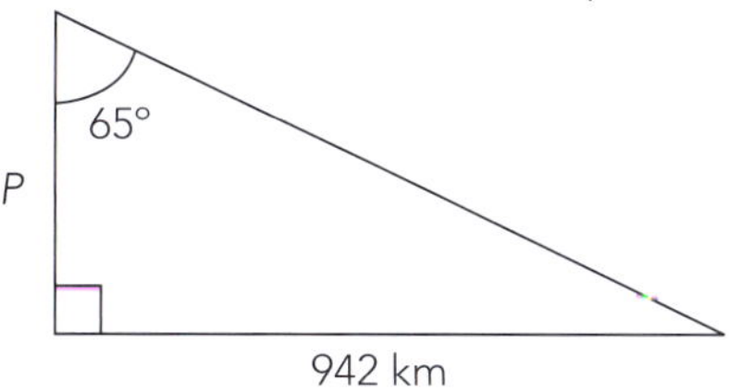

ISBN: 9780170370394

Finding angles

Once again we use the three rules:

$$\sin\theta = \frac{O}{H} \qquad \cos\theta = \frac{A}{H} \qquad \tan\theta = \frac{O}{A}$$

Example: Calculate the angle marked θ.

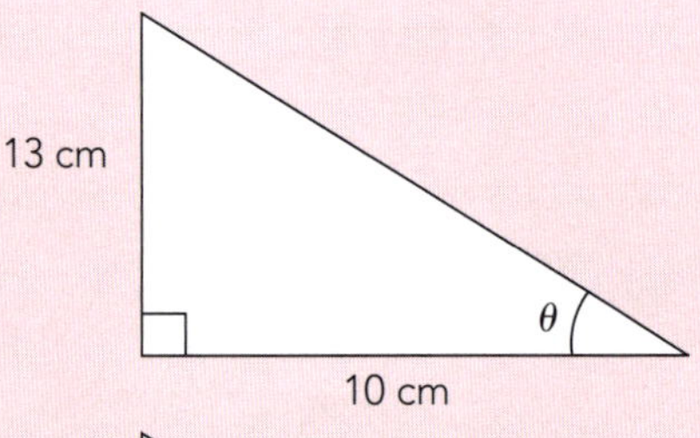

Step 1: **Label** the sides that are involved in the problem with 'A', 'O' and 'H'.

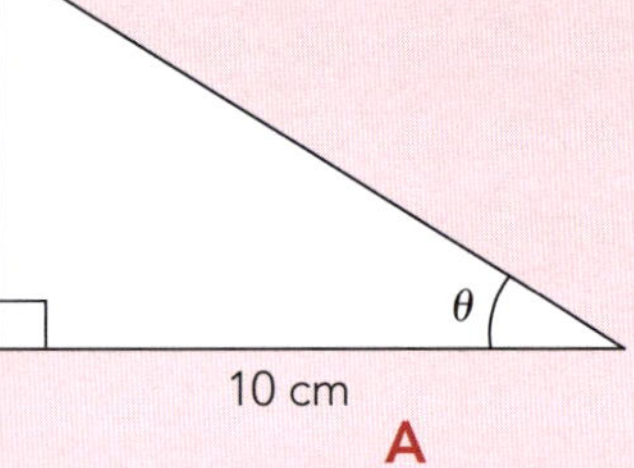

Step 2: Decide **which relationship** involves these sides. In this case we are concerned only with O and A so we must use:

$$\tan\theta = \frac{O}{A}$$

Step 3: **Substitute** the values from the triangle.

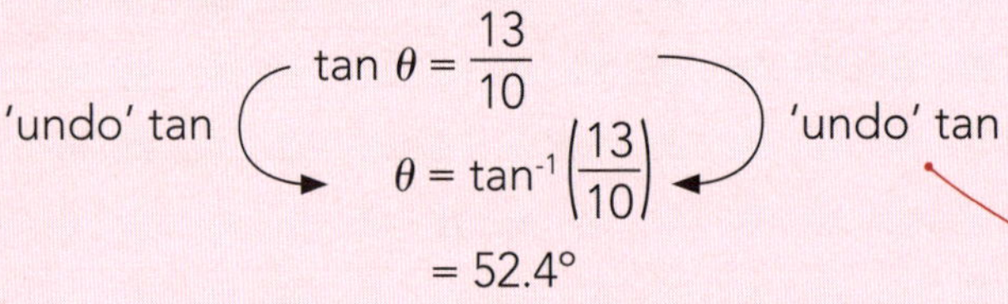

$$\tan\theta = \frac{13}{10}$$

$$\theta = \tan^{-1}\left(\frac{13}{10}\right)$$

$$= 52.4°$$

Use the 'inverse tan' button on your calculator. Don't forget the **brackets**.

Step 4: **Think about your answer — does it seem reasonable?**
In this case, θ must be more than 45° because 13 cm is longer than 10 cm, so an answer of 52.4° is reasonable.

Calculate the angle marked θ in the following triangles.

1

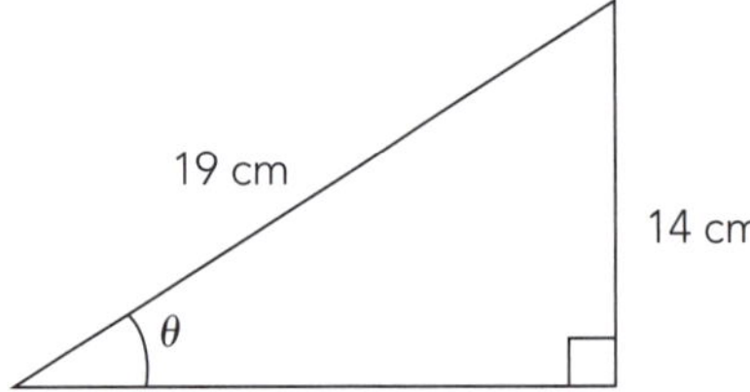

____

____

____

2

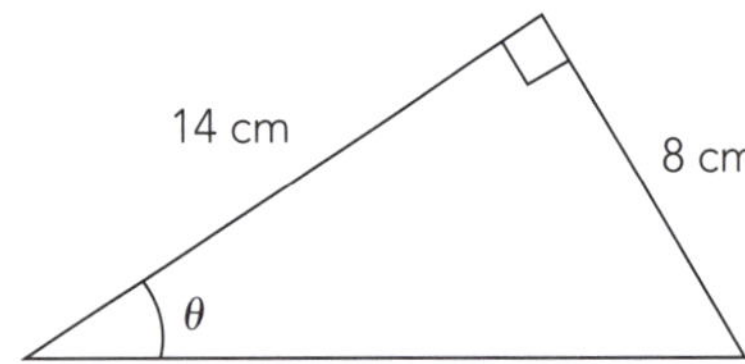

____

____

____

ISBN: 9780170370394

3

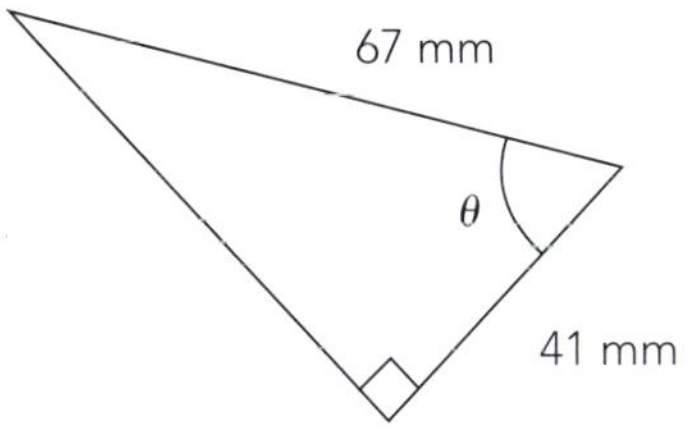

67 mm
θ
41 mm

4

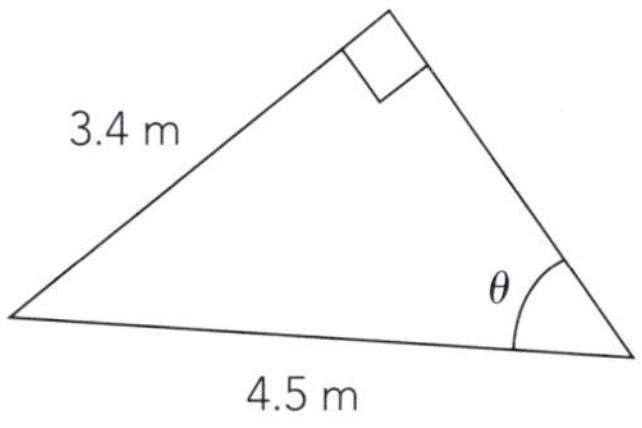

3.4 m
θ
4.5 m

5

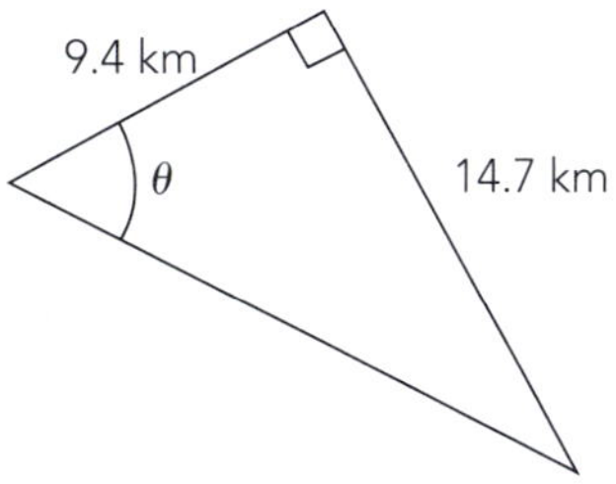

9.4 km
θ
14.7 km

6

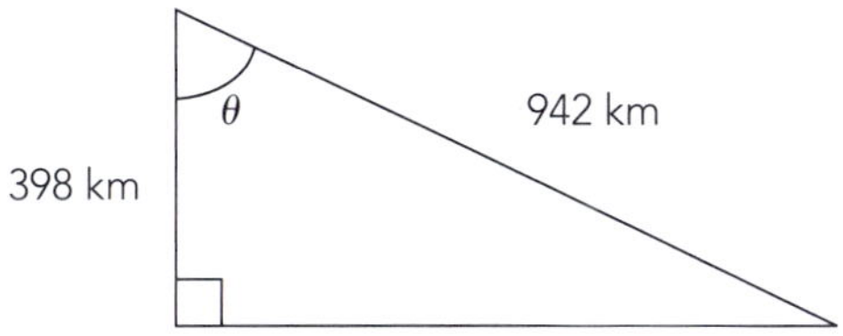

θ
942 km
398 km

7

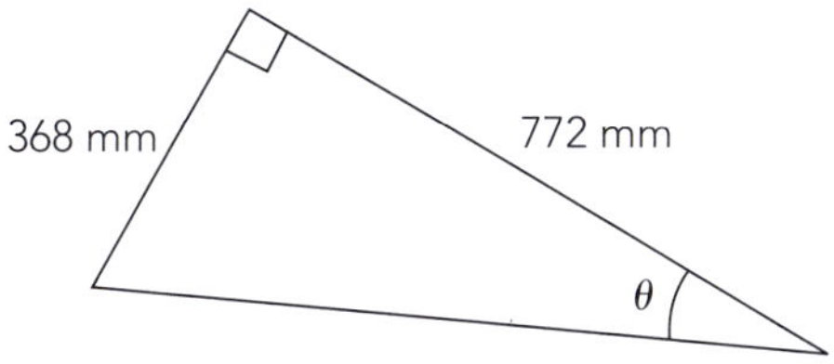

368 mm
772 mm
θ

8

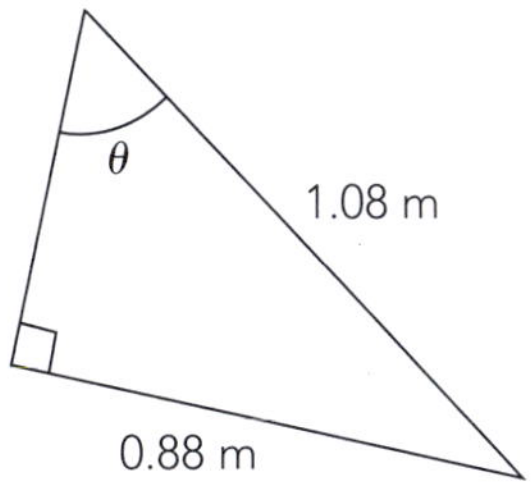

θ
1.08 m
0.88 m

9

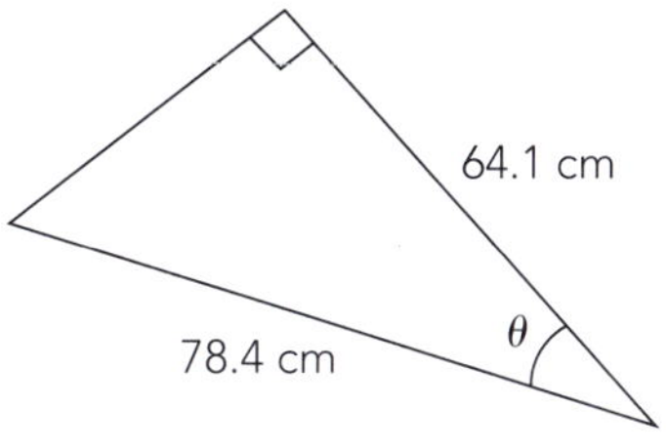

64.1 cm
θ
78.4 cm

ISBN: 9780170370394

Putting it together — with Pythagoras

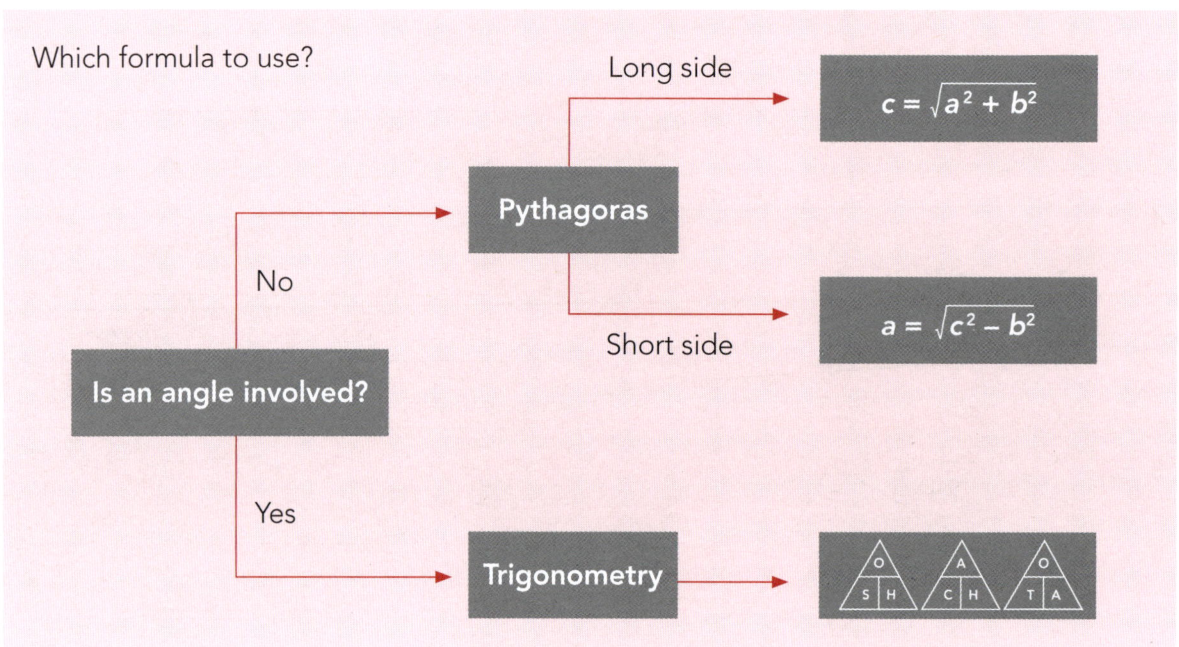

Calculate the unknown sides and angles.

1

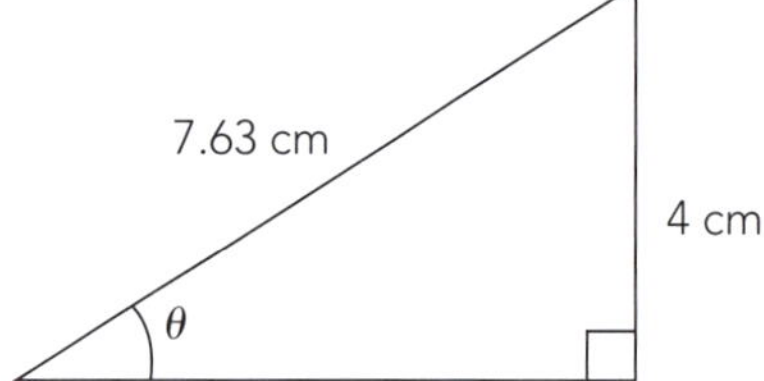

2

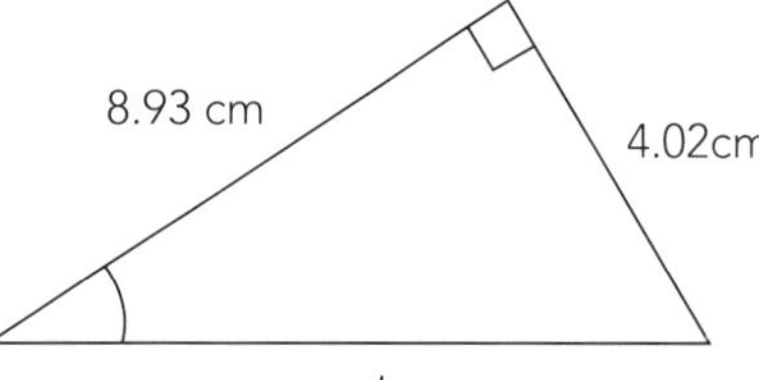

3

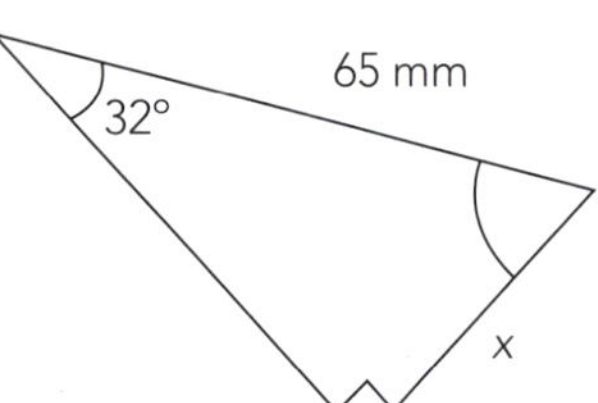

4

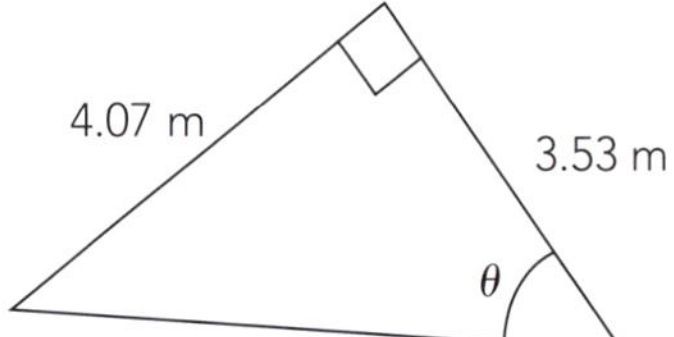

 ISBN: 9780170370394

5 A rule for building ramps is that they may rise only 1 unit for every 12 units of horizontal length. Calculate:

a the angle between the ramp and the ground

b the length of the ramp.

6 Calculate:

a the length *L*

b the angle θ.

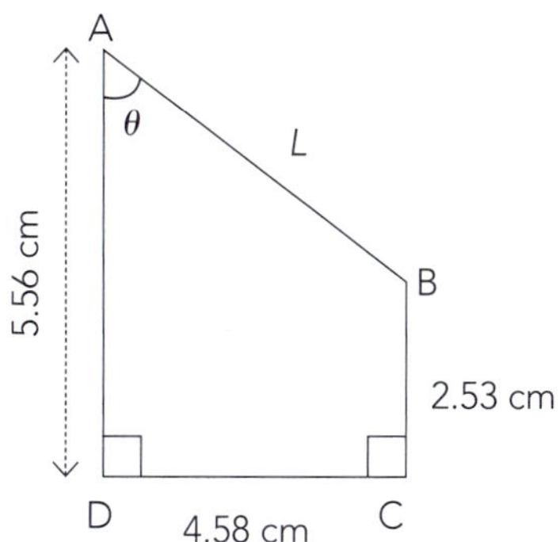

7 Calculate the lengths of:

a *x*

b *y*.

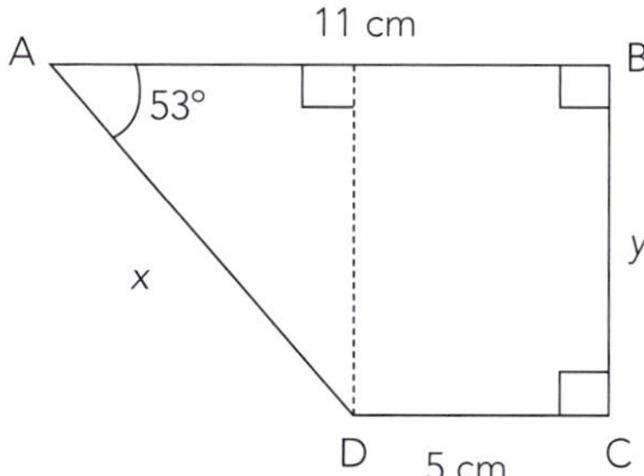

8 ABCD is a rhombus whose diagonals meet at E.

a Calculate θ.

b Calculate the length of diagonal AC.

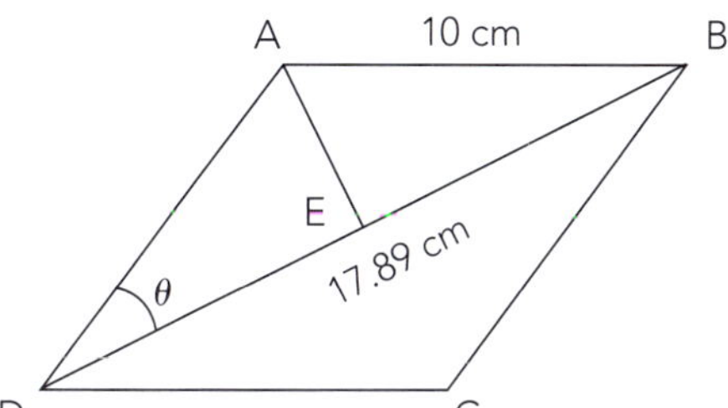

ISBN: 9780170370394

9 Calculate the length of AC.

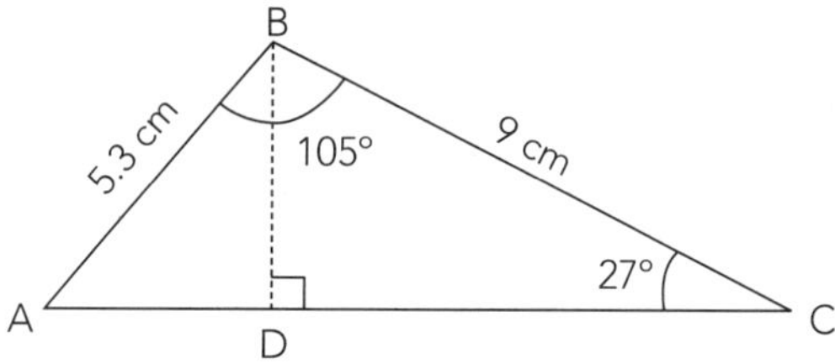

10 A 50 cm square has a triangle cut off it. The base of the triangle is also 50 cm. If AX = CY, calculate h, the height of the triangle.

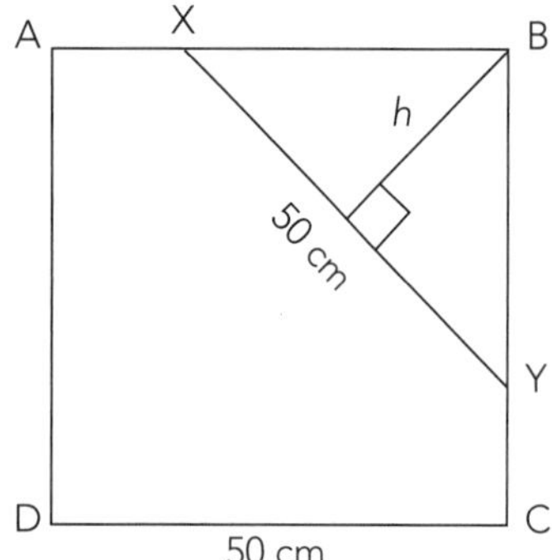

11 ADCE is a kite. The diagonal DE is 8 cm. Calculate the length of the diagonal AC.

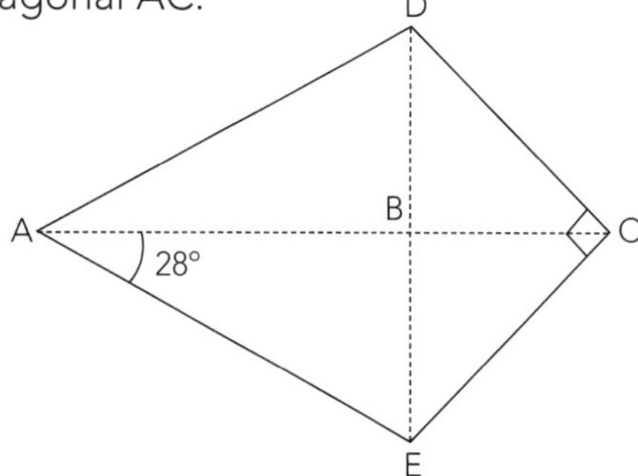

12 Calculate the perimeter of this figure.

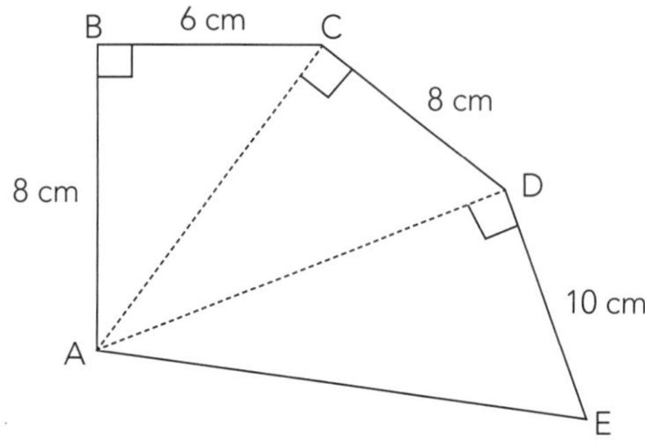

 ISBN: 9780170370394

Angles with lines

Revision

Write a definition and match it to the correct diagram. The first one is done for you.

Term	Description	Diagram
Acute angle	An angle that is less than 90°	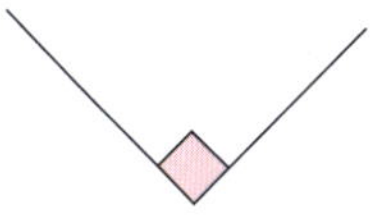
Obtuse angle		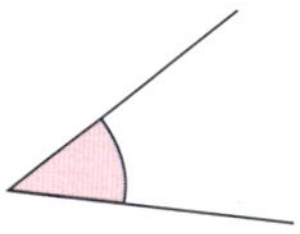
Reflex angle		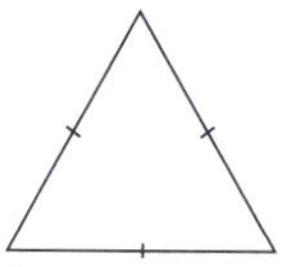
Right angle		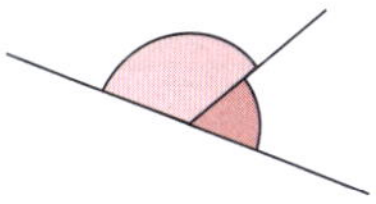
Scalene triangle		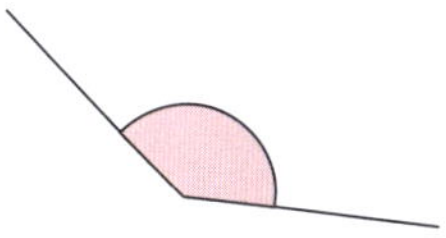
Isosceles triangle		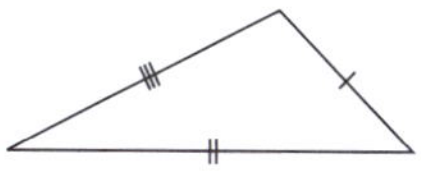
Equilateral triangle		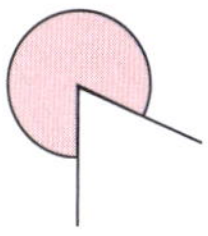
Right-angled triangle		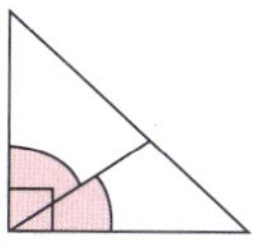
Supplementary angles		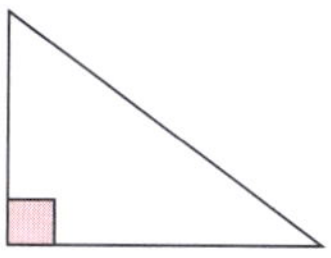
Complementary angles		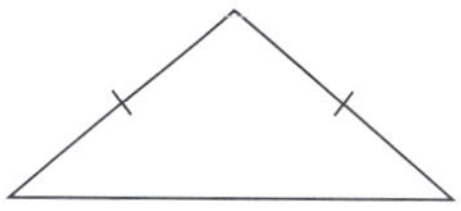

ISBN: 9780170370394

Fundamentals

Relationship	Reason
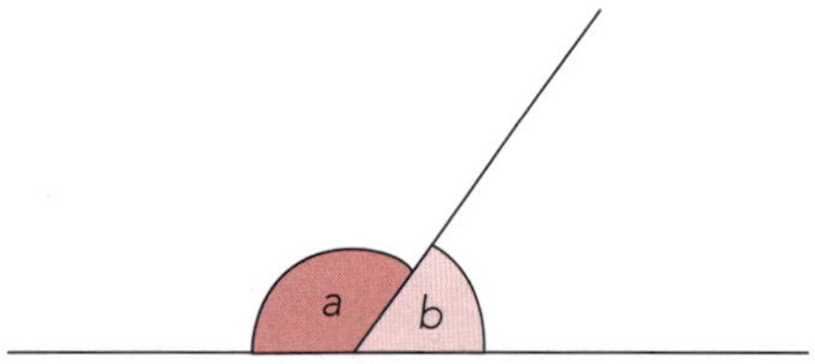 a / b $a + b = 180°$	Angles on a line add to 180°. (∠s on a line = 180°)
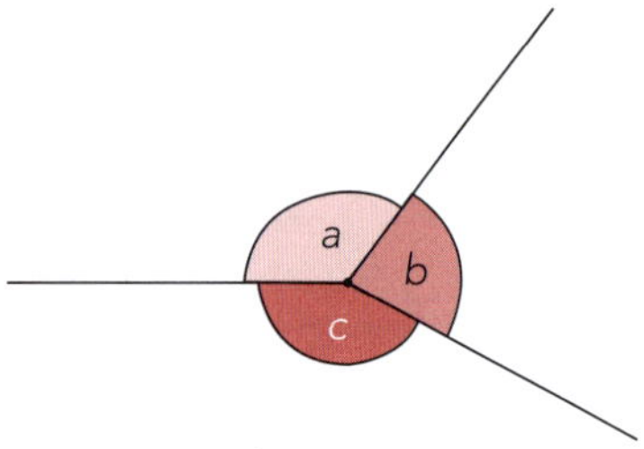 a / b / c $a + b + c = 360°$	Angles at a point add to 360°. (∠s at a point = 360°)
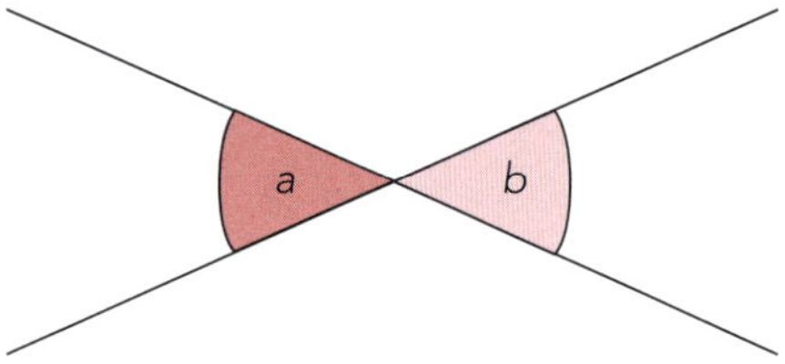 a / b $a = b$	Vertically opposite angles are equal. (vert opp ∠s =)
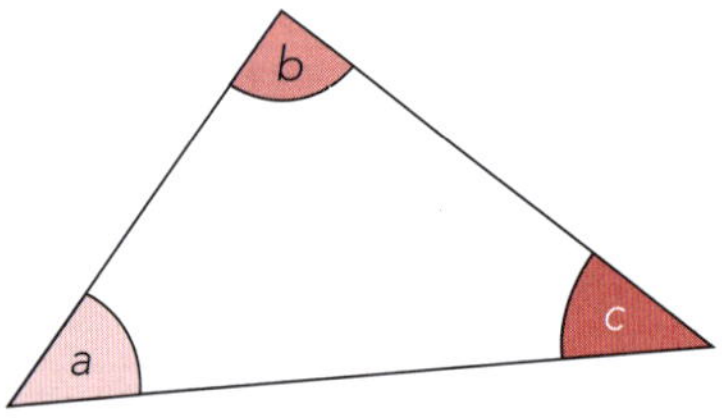 b / a / c $a + b + c = 180°$	Angles in a triangle add to 180°. (∠s in Δ = 180°)
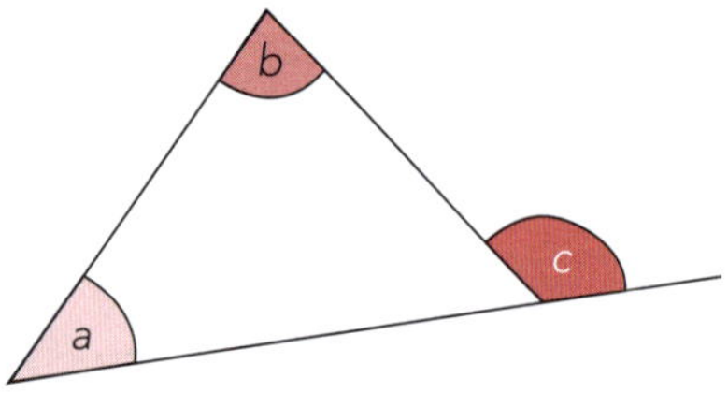 b / c / a $a + b = c$	The exterior angle of a triangle = the sum of the interior opposite angles. (ext ∠ of Δ = sum of int opp ∠s)

ISBN: 9780170370394

Calculate the unknown angles and give reasons.

1

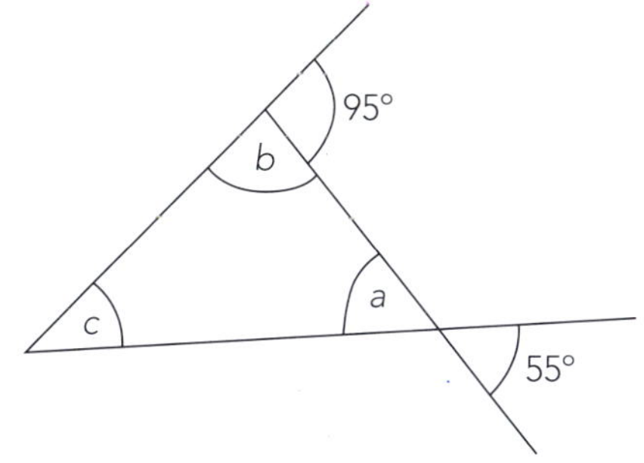

$a =$ ___ ______________________

$b =$ ___ ______________________

$c =$ ___ ______________________

2

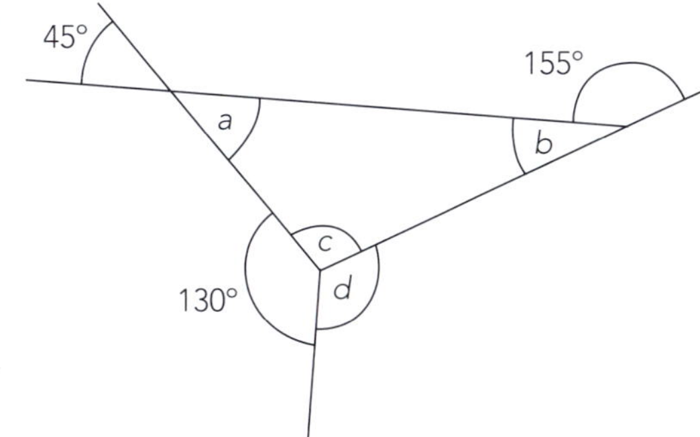

$a =$ ___ ______________________

$b =$ ___ ______________________

$c =$ ___ ______________________

$d =$ ___ ______________________

3

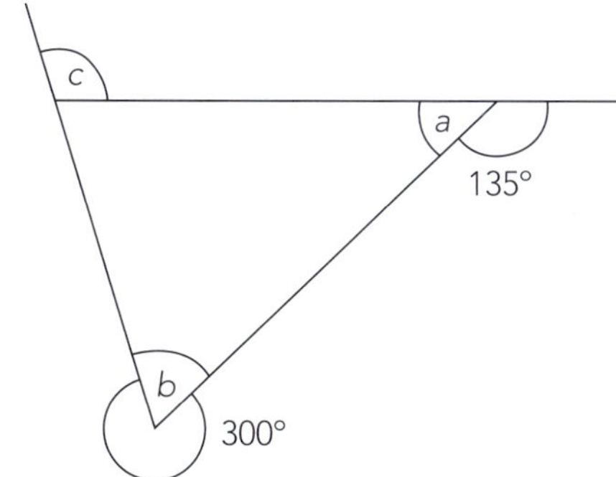

$a =$ ___ ______________________

$b =$ ___ ______________________

$c =$ ___ ______________________

4

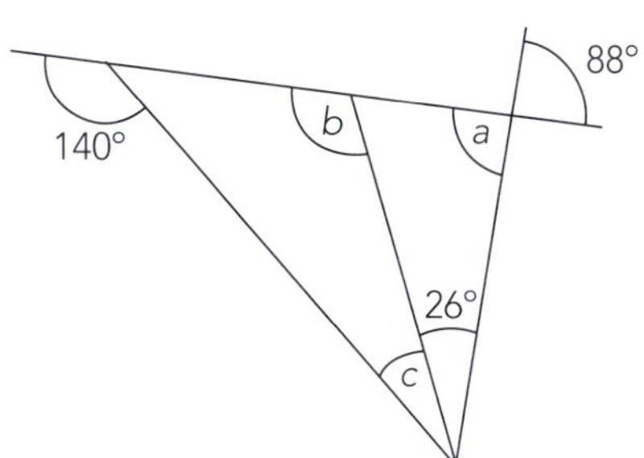

$a =$ ___ ______________________

$b =$ ___ ______________________

$c =$ ___ ______________________

5

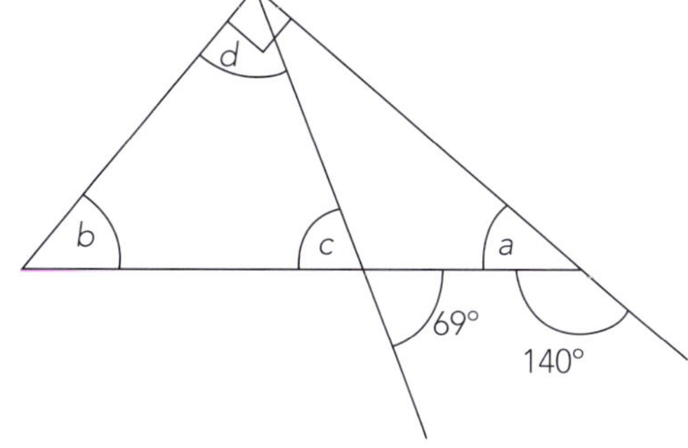

$a =$ ___ ______________________

$b =$ ___ ______________________

$c =$ ___ ______________________

$d =$ ___ ______________________

Naming angles

- Usually, upper-case letters are used to mark ends of lines and vertices.
- Angles are named using three letters.
- The **middle** letter is always **at** the angle.

Example:

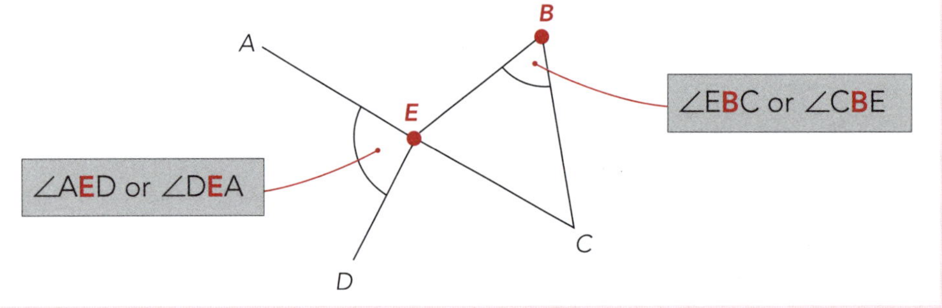

Developing a chain of reasoning

- Often you will need several steps in order to find an answer.
- Begin by finding every angle you can.
- Then fit those needed into a logical sequence that leads from the given information to the answer required.
- You need to give the reasons for each step.
- Often there are several correct solutions, so check with your teacher if yours is different.

Example one:

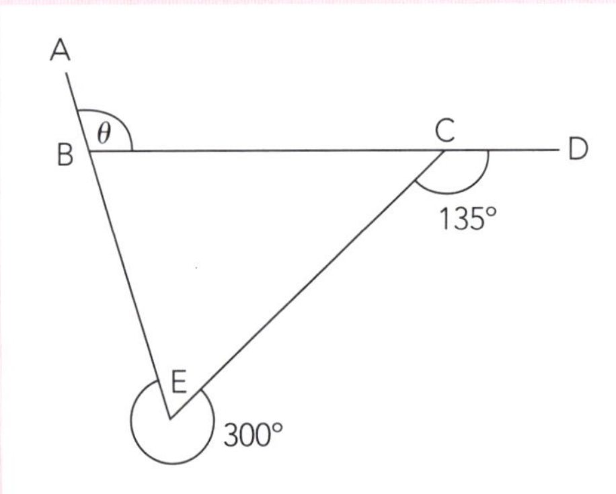

∠BCE = 45° (∠s on a line = 180°)
∠BEC = 60° (∠s at a point = 360°)

∴ θ = 105° (ext ∠ of Δ = sum of int opp ∠s)

Example two:

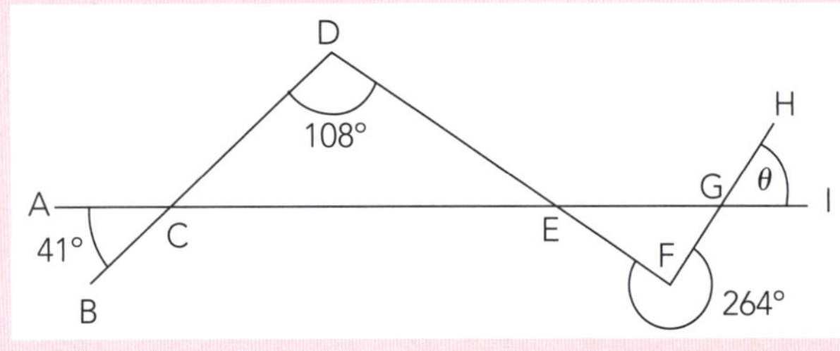

∠DCE = 41° (vert opp ∠s =)
∠DEG = 149° (ext ∠ of Δ = sum of int opp ∠s)
∠EFG = 96° (∠s at a point = 360°)
∠EGF = 149° – 96°
= 53° (ext ∠ of Δ = sum of int opp ∠s)

∴ θ = 53° (vert opp ∠s =)

ISBN: 9780170370394

Calculate the unknown angle and give reasons for each step.

1

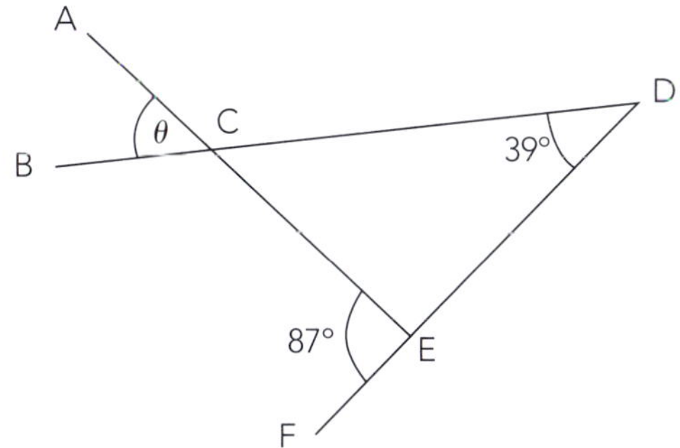

2

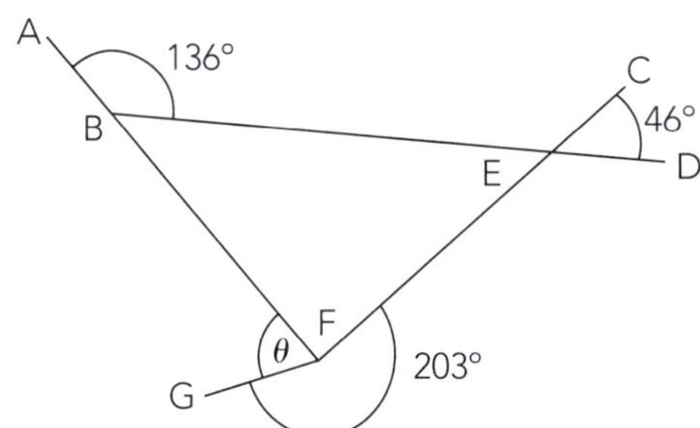

3

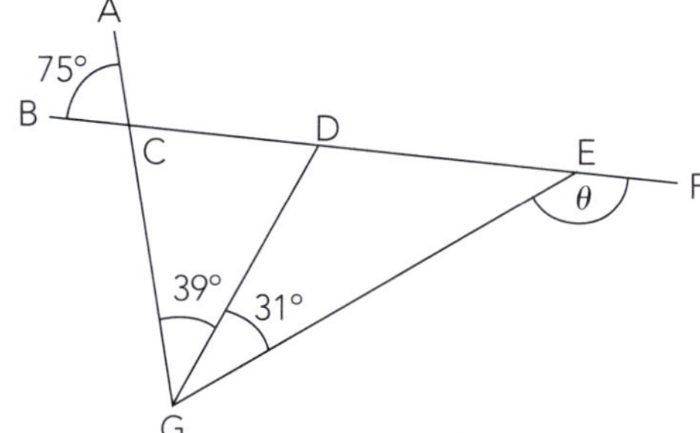

4

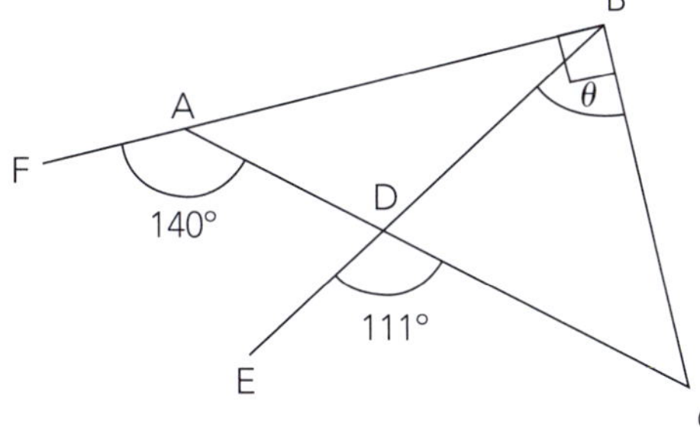

5

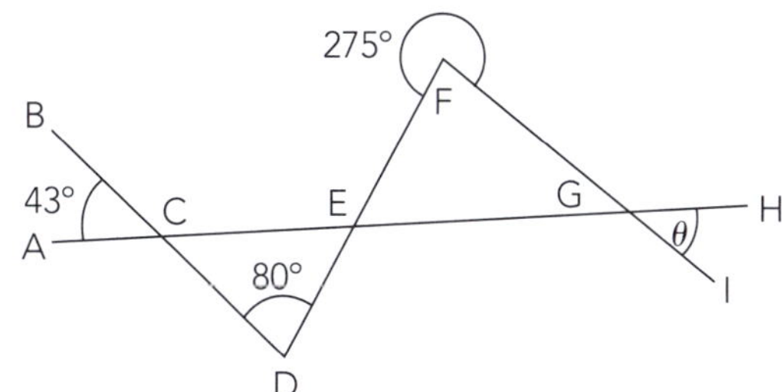

Parallel lines

Relationship	Reason
$a = b$	Alternate angles on parallel lines are equal. (These form a '**Z**'.) (alt ∠s =, // lines)
$a = b$	Corresponding angles on parallel lines are equal. (These form an '**F**'.) (corr ∠s =, // lines)
$a + b = 180°$	Co-interior angles on parallel lines add to 180°. (These form a '**C**'.) (co-int ∠s add to 180°, // lines)

State which of these relationships applies to the following pairs of angles.

1 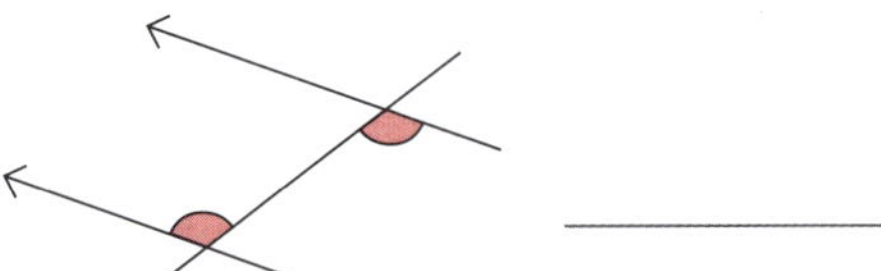 ______________

2 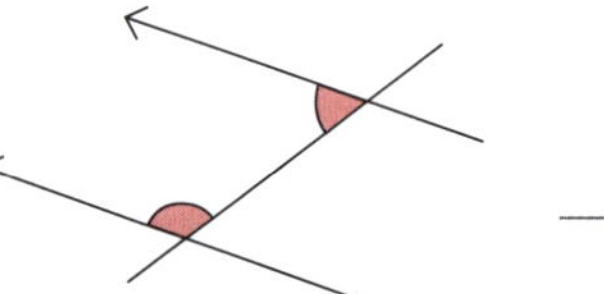 ______________

3 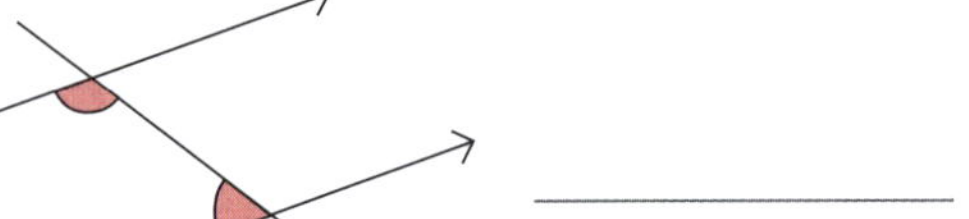 ______________

4 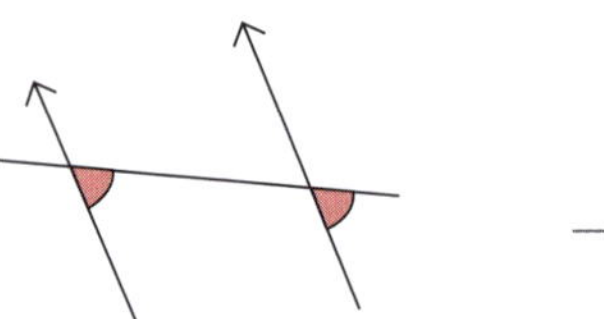 ______________

5 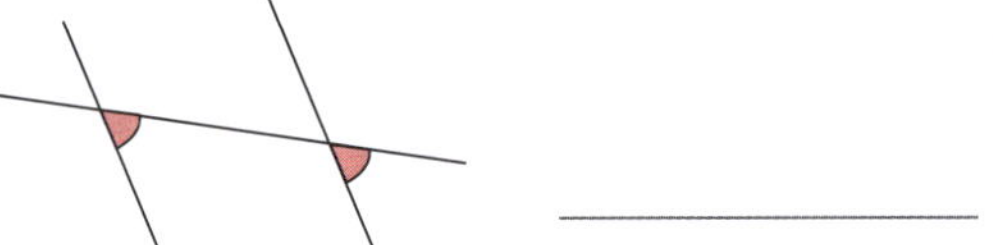 ______________

6 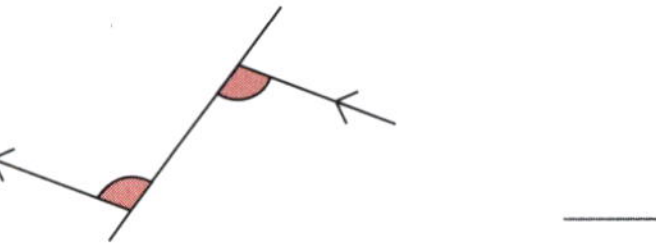 ______________

7 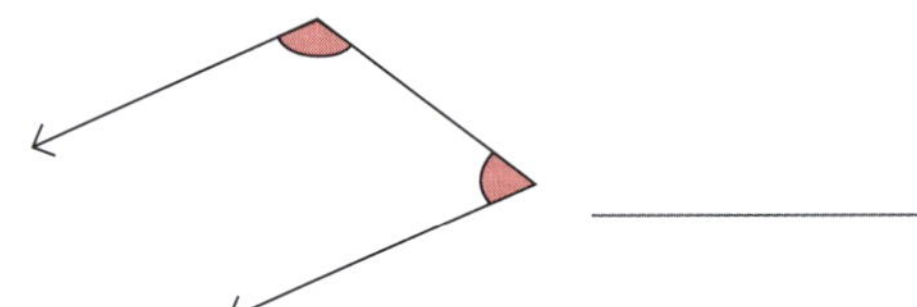 ______________

8 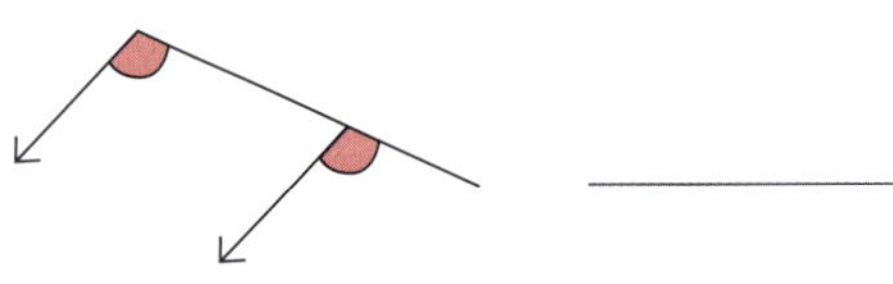 ______________

ISBN: 9780170370394

Calculating angles where parallel lines are involved

- Once again, calculate all the angles you can.
- Be prepared to use the basic angle rules as well as the parallel line rules.
- Sometimes you will need to give two reasons for an angle calculation.

Example:

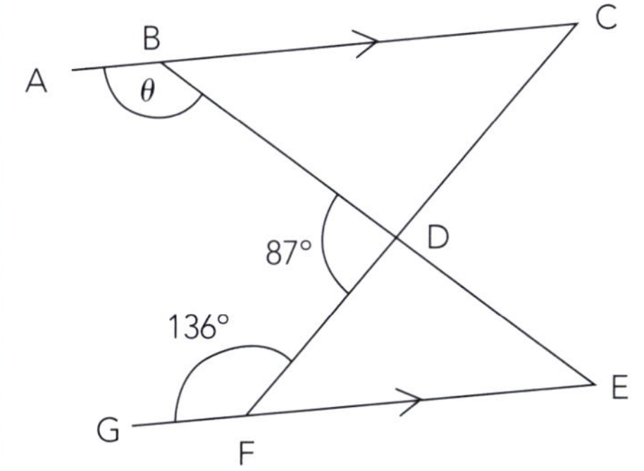

$\angle DCB = 44°$ (co-int $\angle$s add to 180°, // lines)

$\angle BDC = 93°$ ($\angle$s on a line = 180°)

$\therefore \angle ABD = 137°$ (ext $\angle$ of Δ = sum of int opp $\angle$s)

$\theta = 137°$

Calculate the unknown angles and give reasons for each step.

1

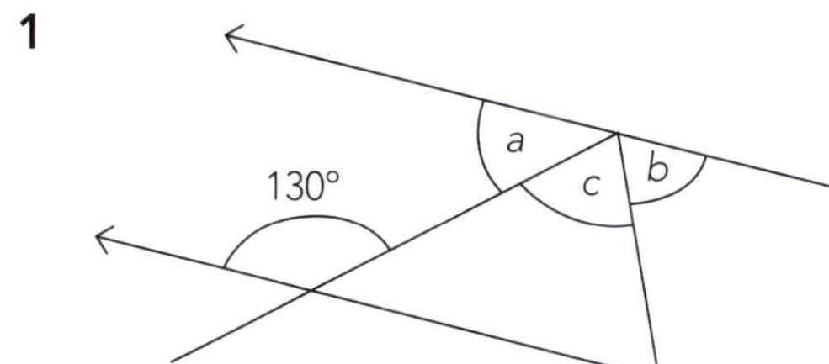

a = ___ ________________________

b = ___ ________________________

c = ___ ________________________

2

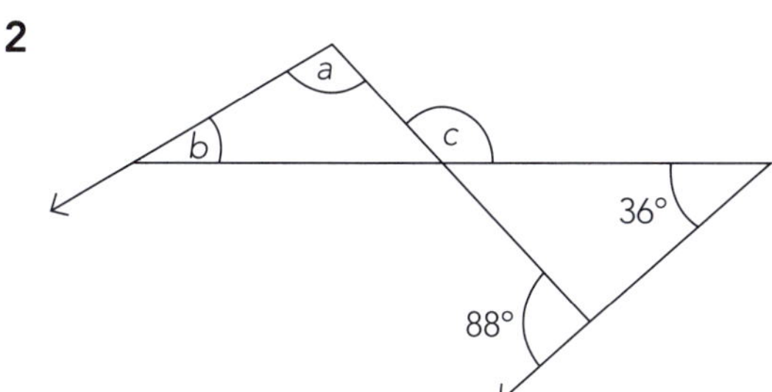

a = ___ ________________________

b = ___ ________________________

c = ___ ________________________

3

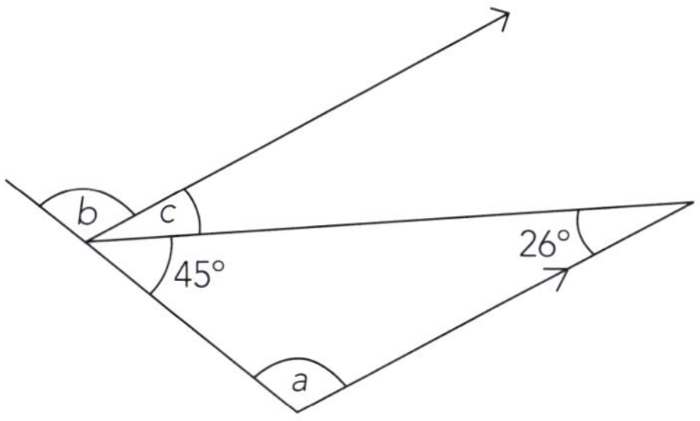

a = ___ ________________________

b = ___ ________________________

c = ___ ________________________

ISBN: 9780170370394

4

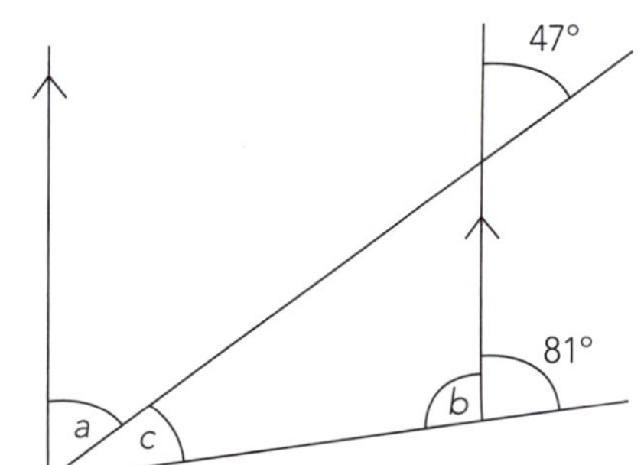

$a =$ ___ ________________________

$b =$ ___ ________________________

$c =$ ___ ________________________

5

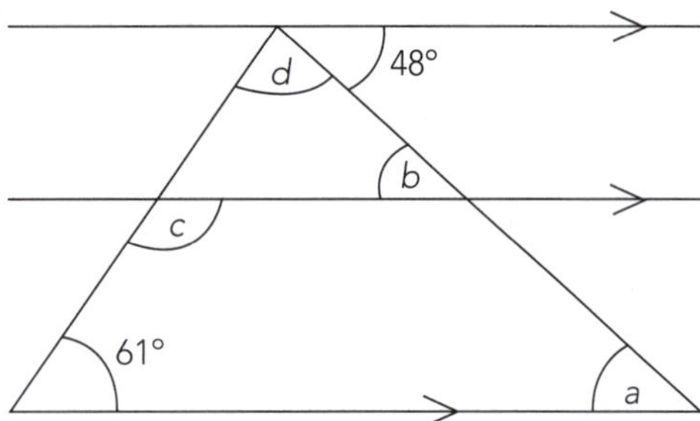

$a =$ ___ ________________________

$b =$ ___ ________________________

$c =$ ___ ________________________

$d =$ ___ ________________________

6

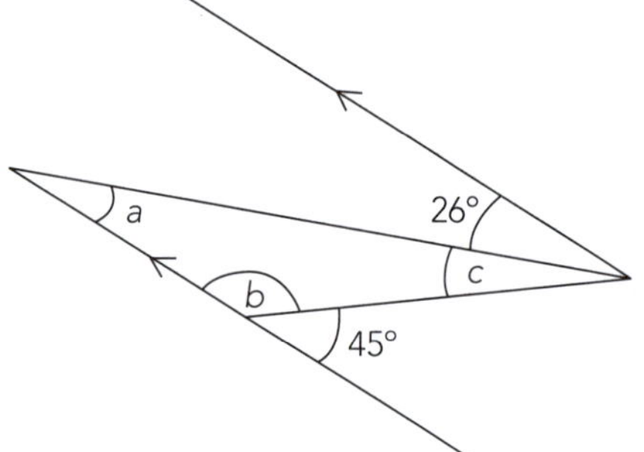

$a =$ ___ ________________________

$b =$ ___ ________________________

$c =$ ___ ________________________

7

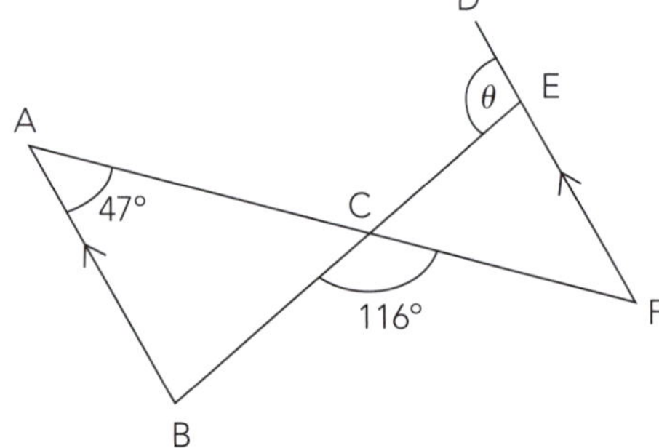

____

____

____

____

8

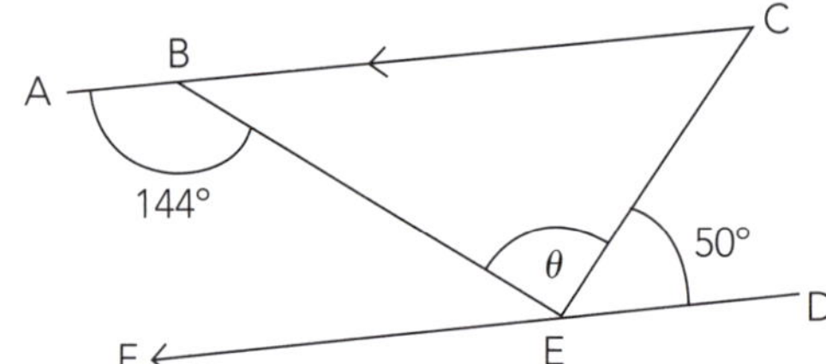

____

____

____

____

ISBN: 9780170370394

9

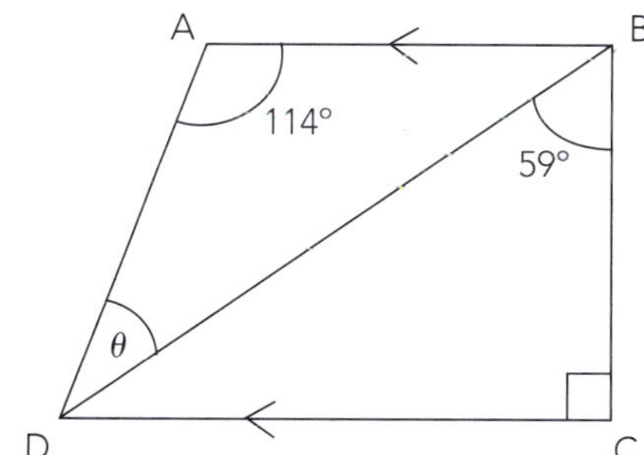

10

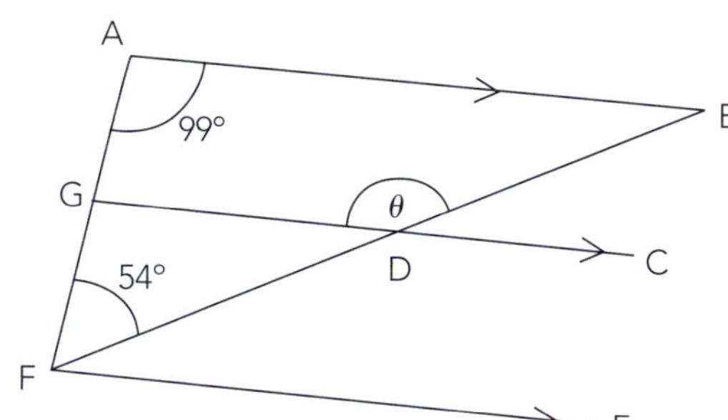

11

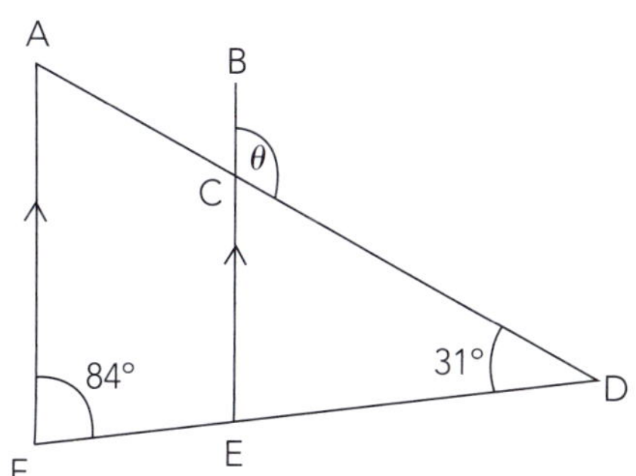

12

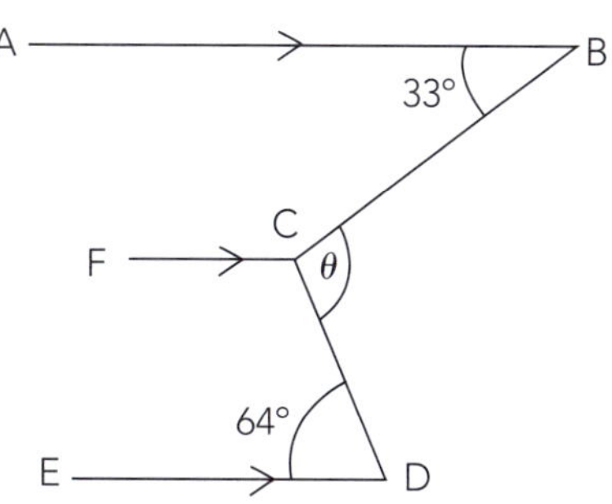

13

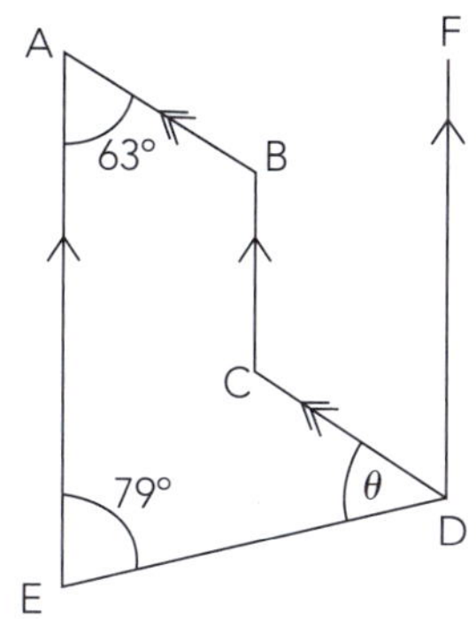

ISBN: 9780170370394

Polygons

Triangles

- Internal angles add to 180°.

Remember:

1 For all triangles

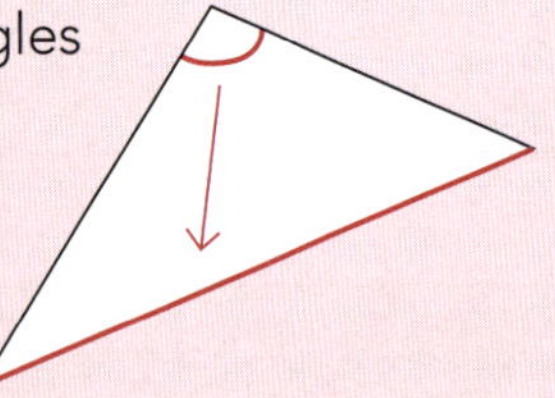

and

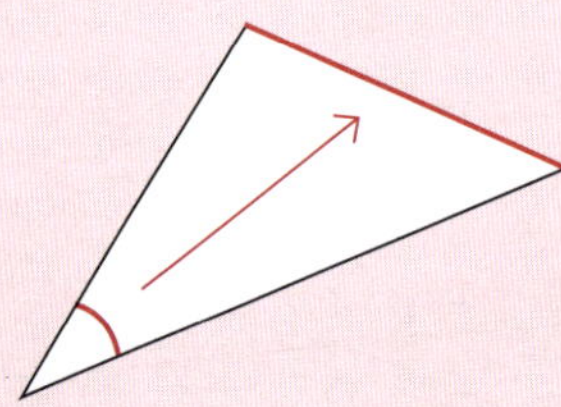

The biggest side is opposite the biggest angle.

The smallest side is opposite the smallest angle.

2 Isosceles triangles

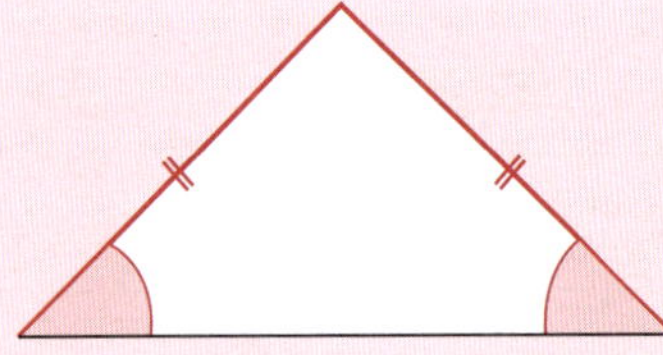

Base angles of an isosceles triangle are equal.

(isos Δ, base ∠s =)

3 Equilateral triangles

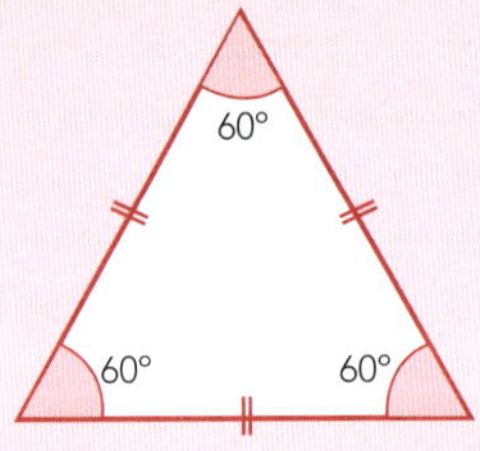

All angles = 60°.

(equilat Δ)

Calculate the unknown angles and give reasons for each step.

1

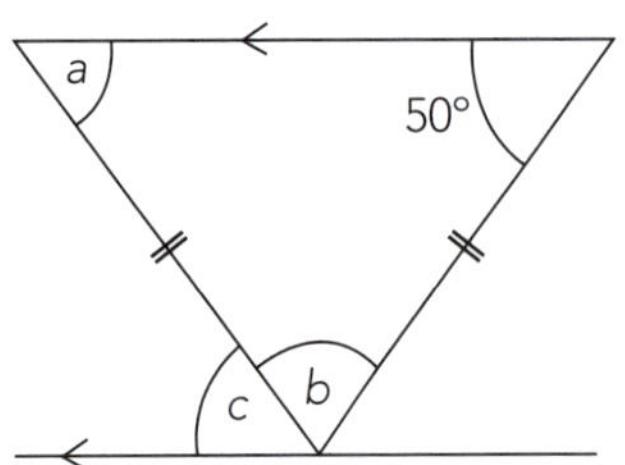

$a =$ ___ ______________________

$b =$ ___ ______________________

$c =$ ___ ______________________

2

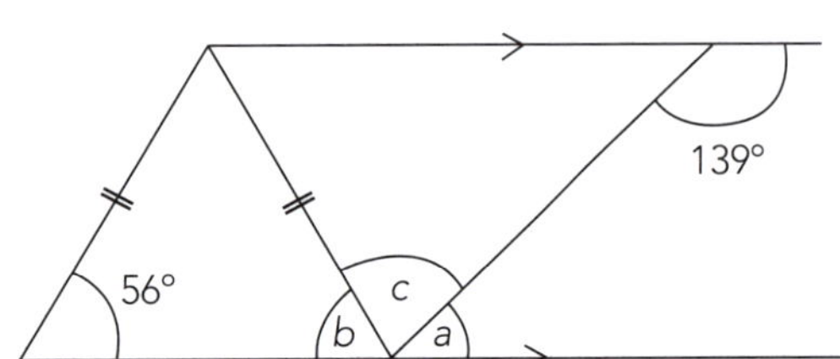

$a =$ ___ ______________________

$b =$ ___ ______________________

$c =$ ___ ______________________

ISBN: 9780170370394

3

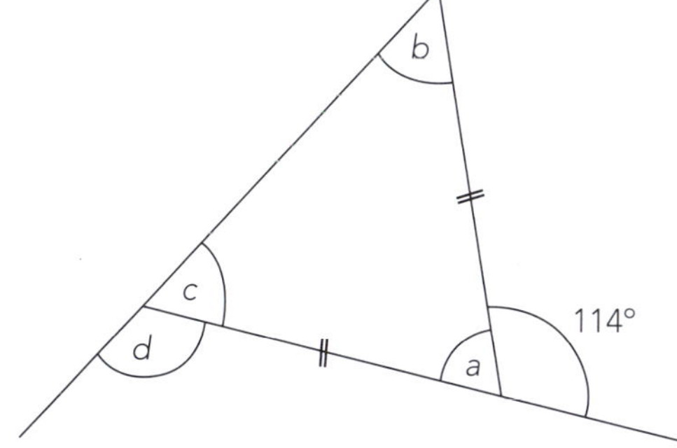

$a =$ ___ ______________________

$b =$ ___ ______________________

$c =$ ___ ______________________

$d =$ ___ ______________________

4

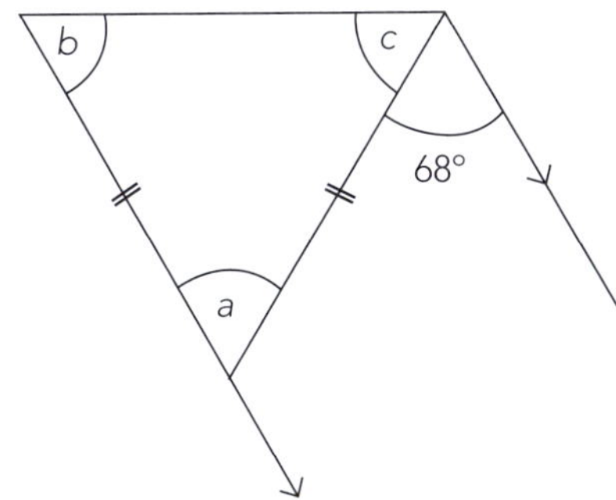

$a =$ ___ ______________________

$b =$ ___ ______________________

$c =$ ___ ______________________

5

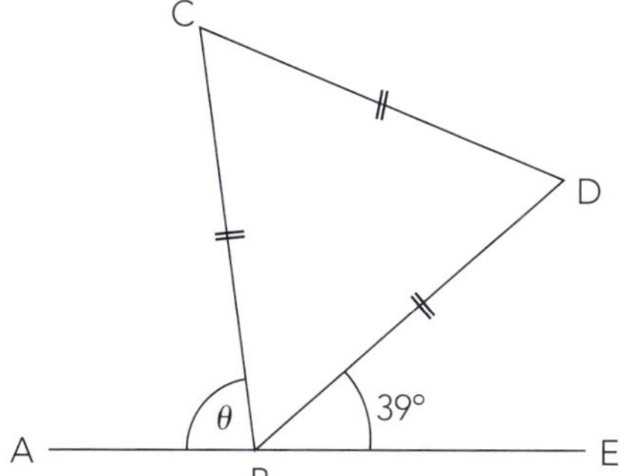

____

____

____

____

6

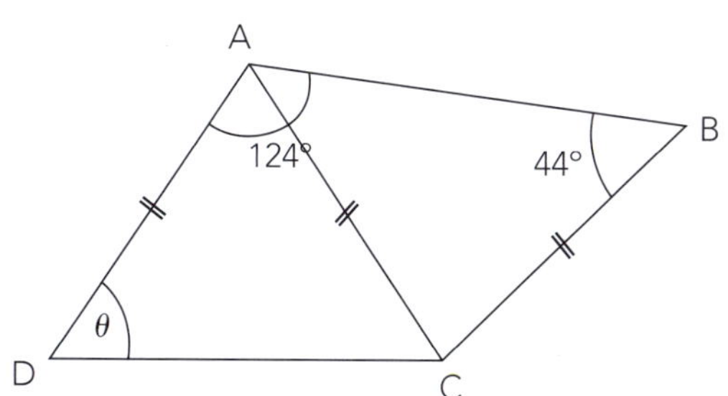

____

____

____

____

7

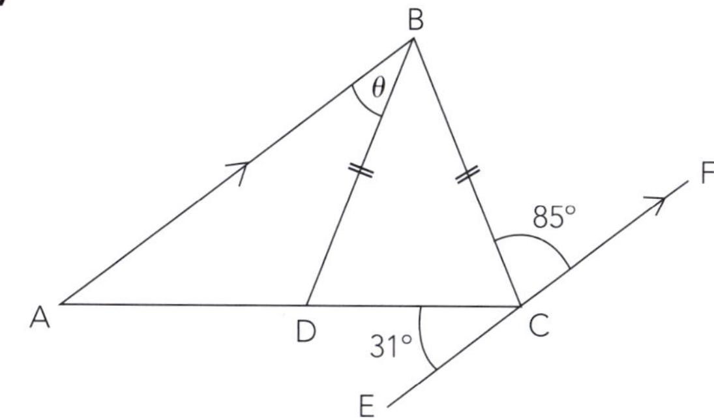

____

____

____

____

Quadrilaterals

- Internal angles add to 360°.

Properties of special quadrilaterals — complete the table

Name	Diagram	Angles	Side lengths	Parallel sides	Diagonals
Square		Right angles at each vertex	Four equal	Two pairs, opposite each other	Equal, perpendicular, bisect each other
Rhombus			Four equal		Not equal, perpendicular, bisect each other
Rectangle					Equal, not perpendicular, bisect each other
Parallelogram			Opposite sides equal, adjacent sides not	Two pairs, opposite each other	
Kite			Two pairs of equal sides, adjacent to each other		Not equal, perpendicular, one is bisected by the other
Trapezium		No right angles, no equal angles	None equal		
Isosceles trapezium		Two pairs of equal angles adjacent to each other			Equal, not perpendicular, do not bisect each other

ISBN: 9780170370394

Calculate the unknown angles and give reasons for each step.

1

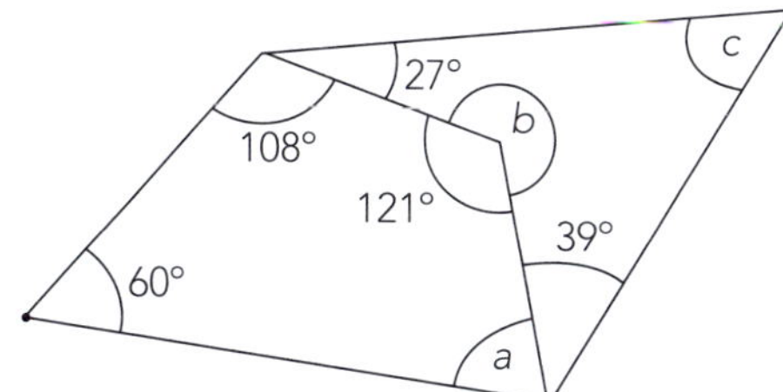

$a =$ ___ ______________________________

$b =$ ___ ______________________________

$c =$ ___ ______________________________

2

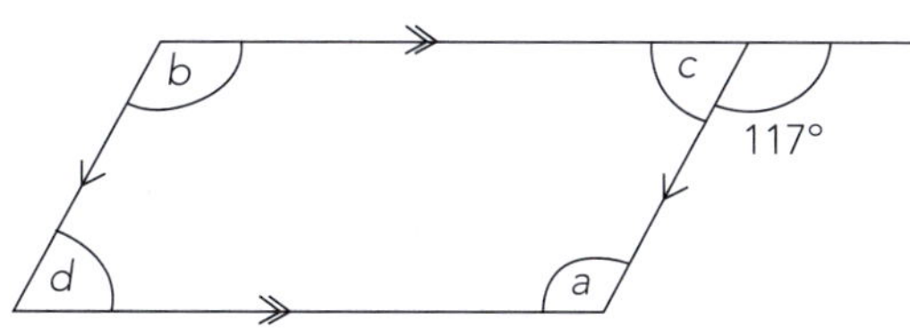

$a =$ ___ ______________________________

$b =$ ___ ______________________________

$c =$ ___ ______________________________

$d =$ ___ ______________________________

3

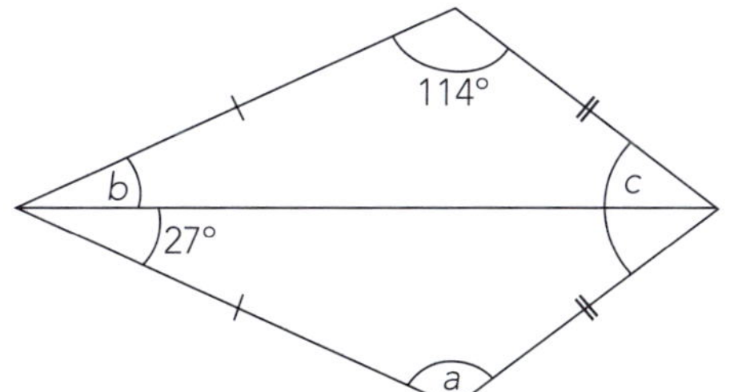

$a =$ ___ ______________________________

$b =$ ___ ______________________________

$c =$ ___ ______________________________

4

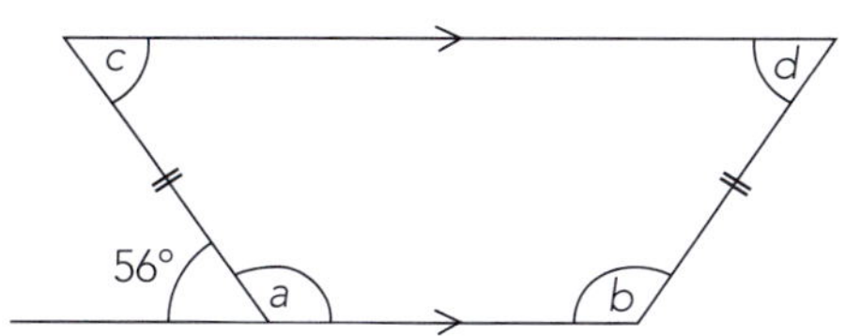

$a =$ ___ ______________________________

$b =$ ___ ______________________________

$c =$ ___ ______________________________

$d =$ ___ ______________________________

5

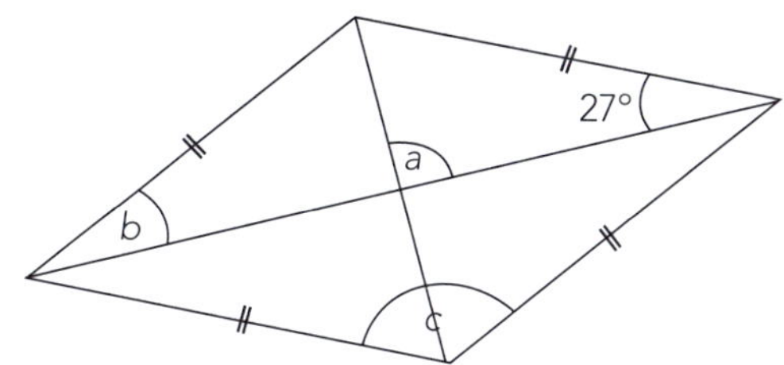

$a =$ ___ ______________________________

$b =$ ___ ______________________________

$c =$ ___ ______________________________

6

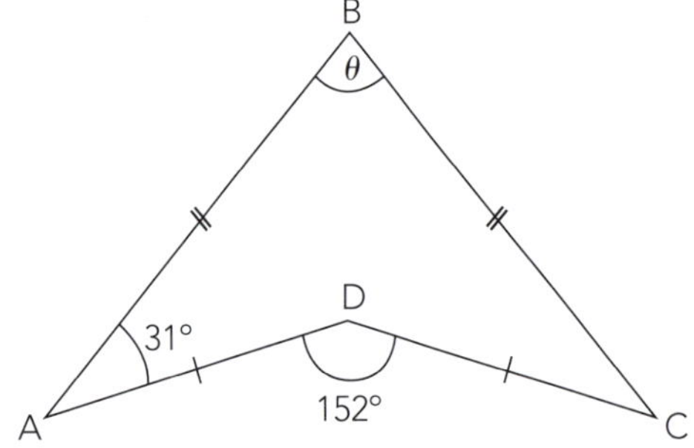

7

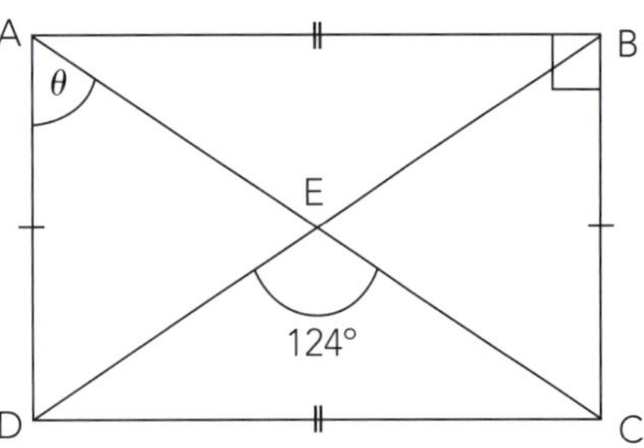

8

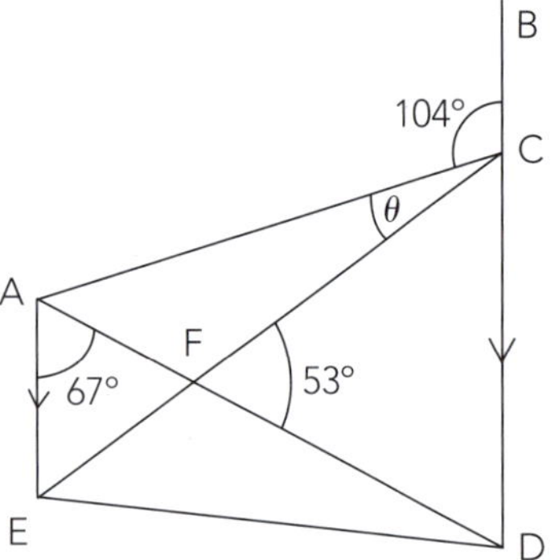

9

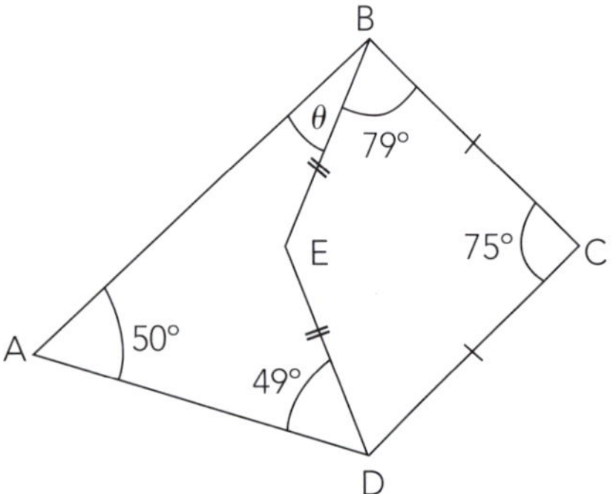

10

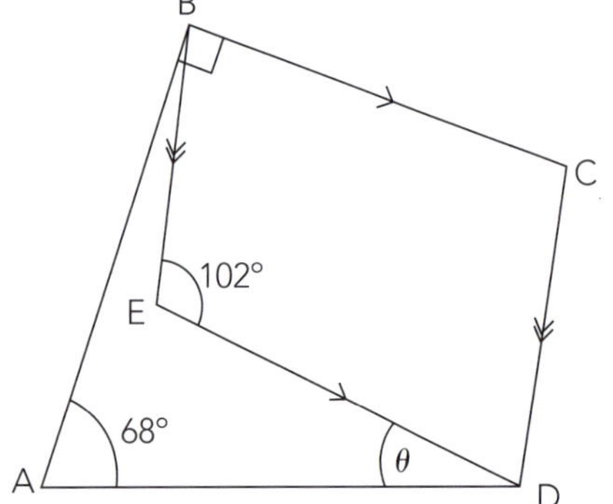

ISBN: 9780170370394

Polygons in general

- Exterior and interior angles

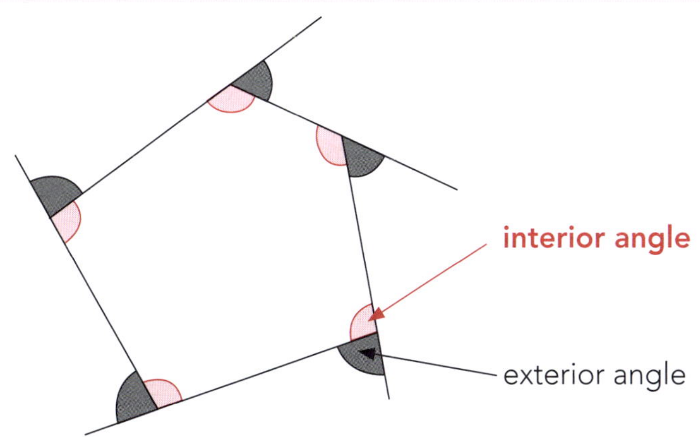

- Regular and irregular polygons

Regular polygons
— have equal sides and angles

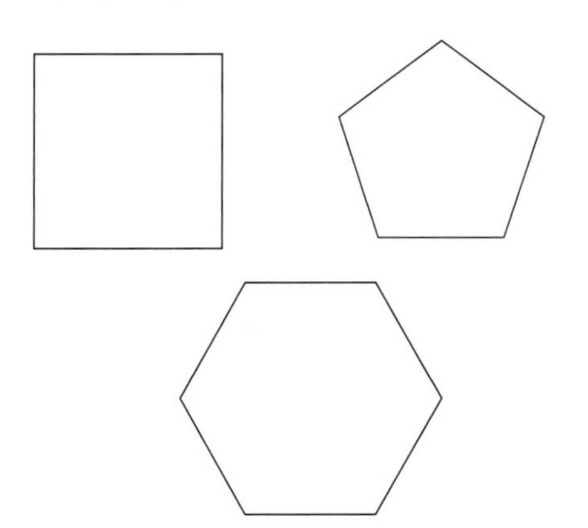

Irregular polygons
— do not have equal sides and angles

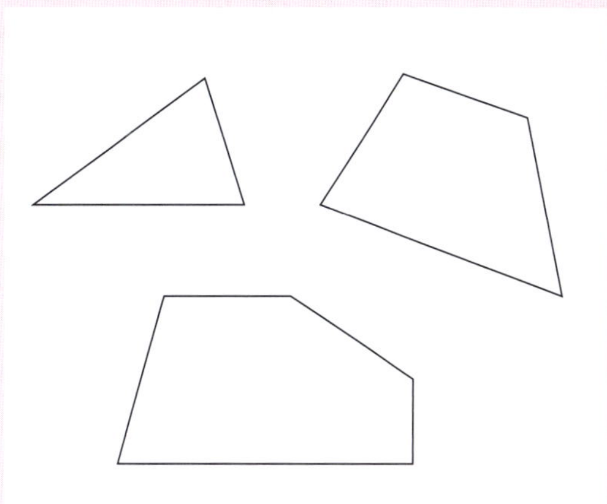

- Names

Name	Number of sides
triangle	3
quadrilateral	4
pentagon	5
hexagon	6
heptagon	7
octagon	8
nonagon	9
decagon	10

ISBN: 9780170370394

Exterior angles of polygons

- These always add to 360° — the following diagrams demonstrate this:

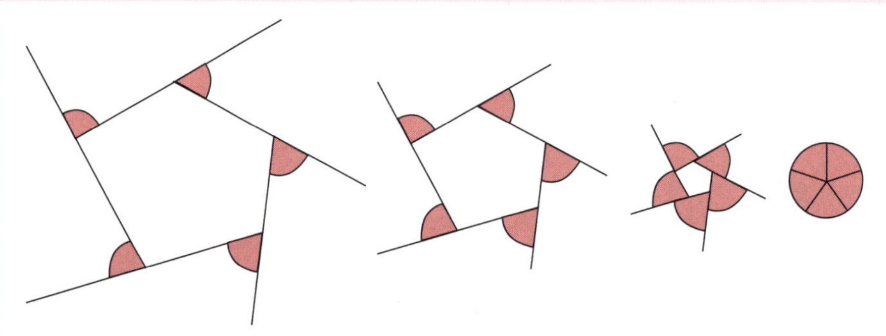

Interior angles of polygons

- Calculate the sum of the interior angles by dividing the polygon into triangles.
- You know the interior angles of a triangle add to 180°.

Name	Number of sides	Diagram	Sum of interior angles	Size of each interior angle of a *regular* polygon
Quadrilateral	**4**	1 2	2 x 180° = 360°	$\frac{360°}{4} = 90°$
Pentagon	**5**	1 2 3	3 x 180° = 540°	$\frac{540°}{5} = 108°$
Hexagon	**6**	1 2 3 4	4 x 180° = 720°	$\frac{720°}{6} = 120°$
Heptagon	**7**	1 2 3 4 5	5 x 180° = 900°	$\frac{900°}{7} = 128.6°$
Octagon	**8**	1 2 3 4 5 6	6 x 180° = 1080°	$\frac{1080°}{8} = 135°$
Any polygon	***n***	The number of triangles is always 2 fewer than the number of sides.	**$(n - 2) \times 180°$**	**$\frac{(n-2) \times 180°}{n}$**

 ISBN: 9780170370394

Exterior and interior angles of polygons

Example one:
The diagram shows a regular octagon.
Calculate the size of θ.

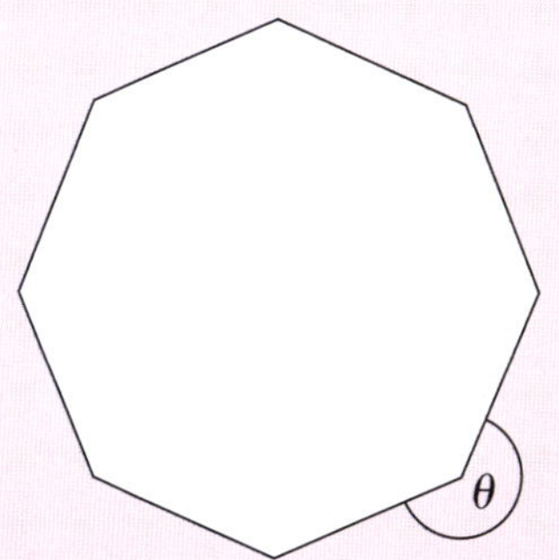

Each exterior angle = $\frac{360°}{8}$ = 45°

$\therefore$ θ = 180° + 45°
= 225°

Example two:
Calculate the size of θ.

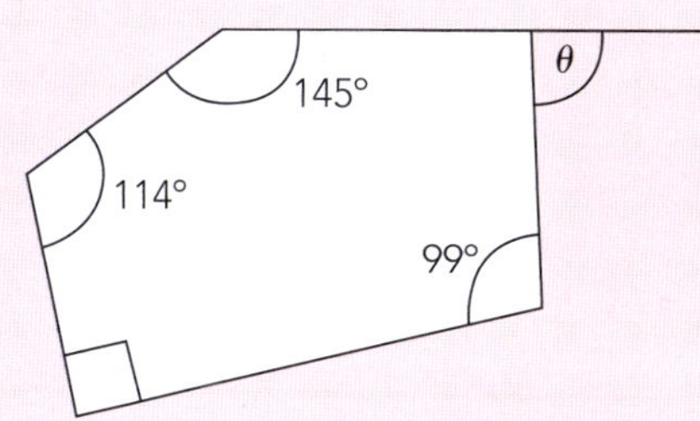

The interior angles of a pentagon add to 3 x 180° = 540°.

$\therefore$ 99° + 90° + 114° + 145° + (180° – θ) = 540°
(∠s on a line = 180°)

θ = 99° + 90° + 114° + 145° + 180° – 540°
= 88°

Calculate the unknown angles.

1

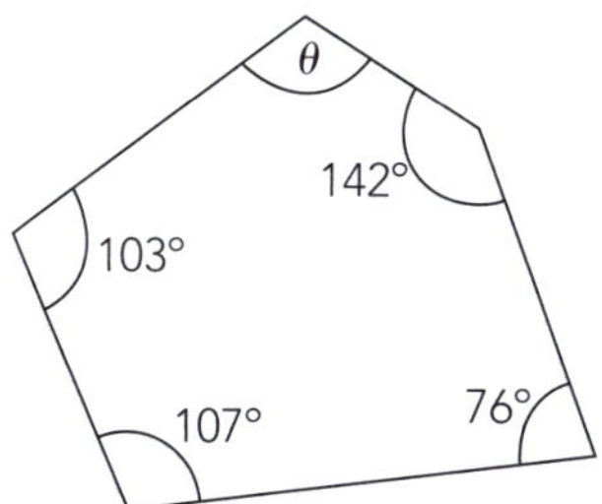

2

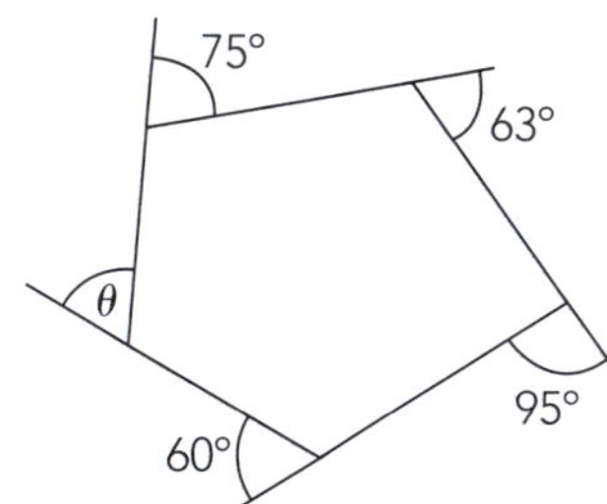

3

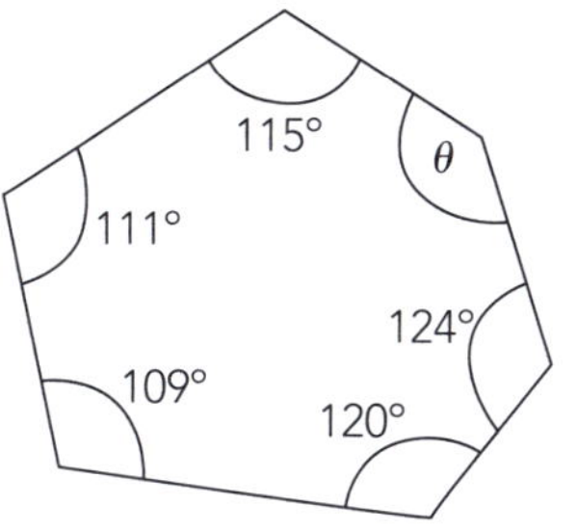

4

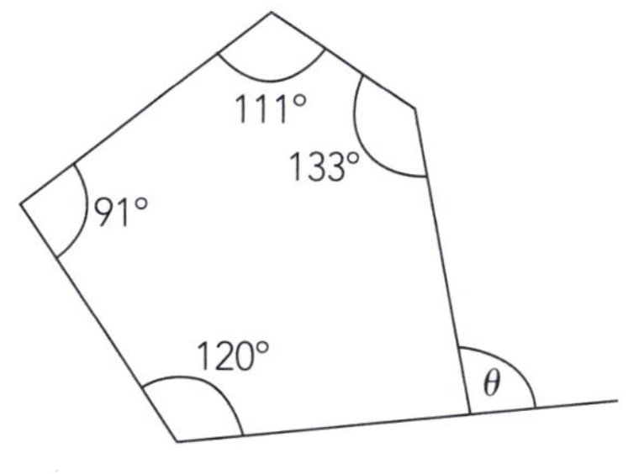

ISBN: 9780170370394

5

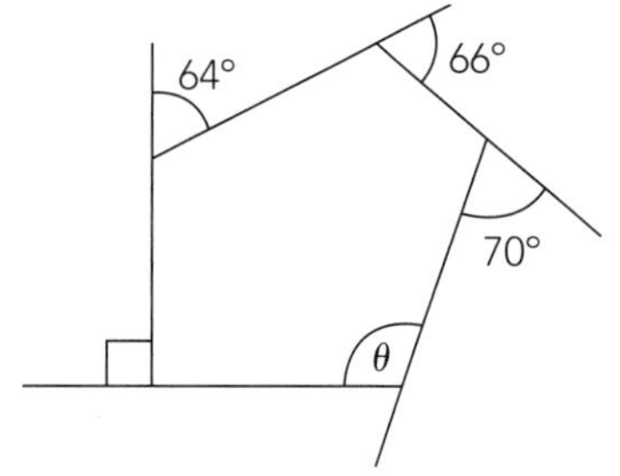

6

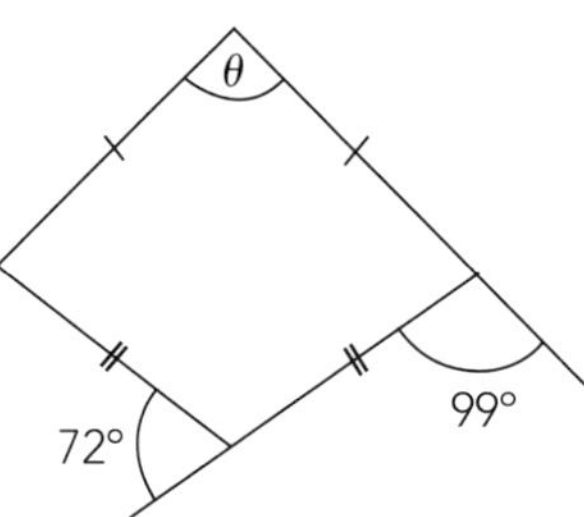

7

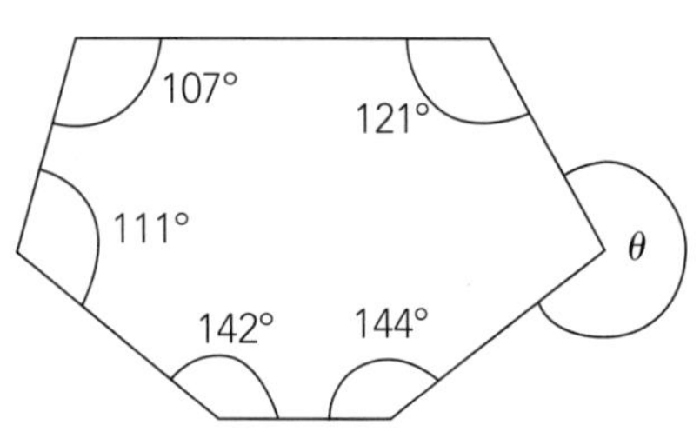

8

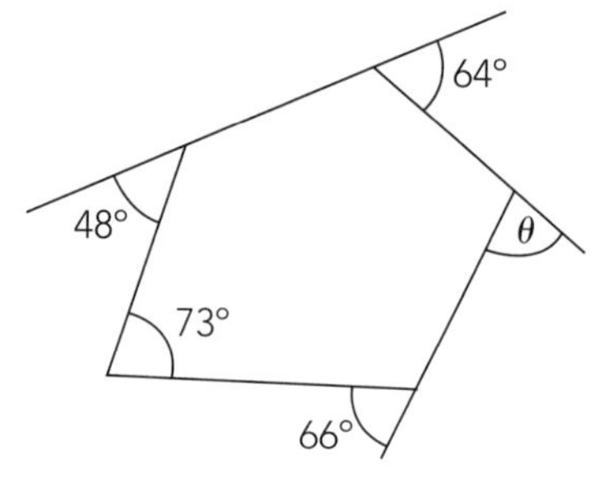

9

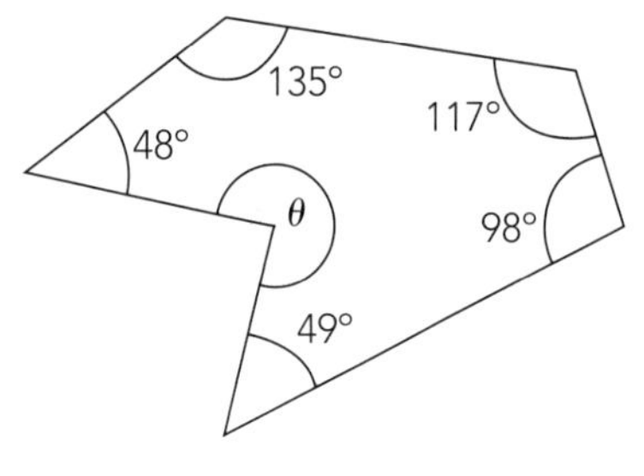

10

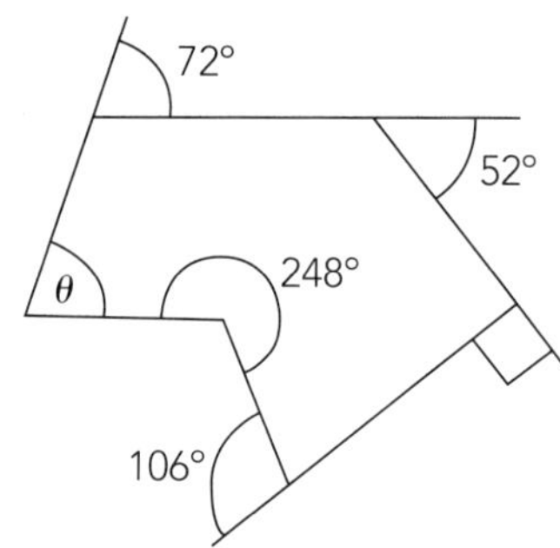

11 Calculate the size of each external angle of a regular decagon.

12 Calculate the size of each internal angle of a regular nonagon.

ISBN: 9780170370394

3D shapes

- When asked to find the length of a side or an angle in a 3D shape, it is a good idea to draw the 2D figure within which it lies.
- These usually involve right-angled triangles, so you can expect to use the Theorem of Pythagoras and trigonometry.

Example: This figure represents a cuboid.

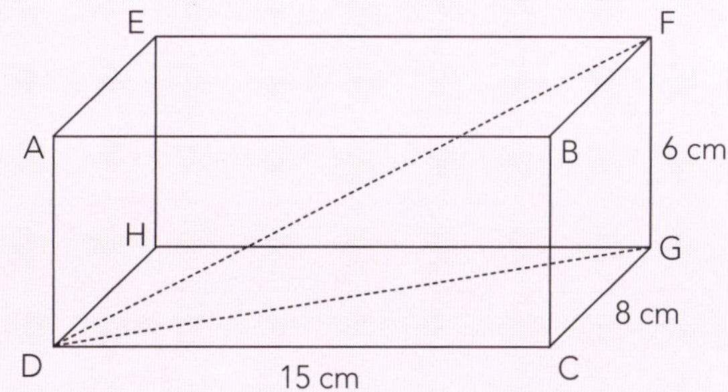

1 Calculate the length of DG.
DG lies in a triangle on the 'floor' of the cuboid.

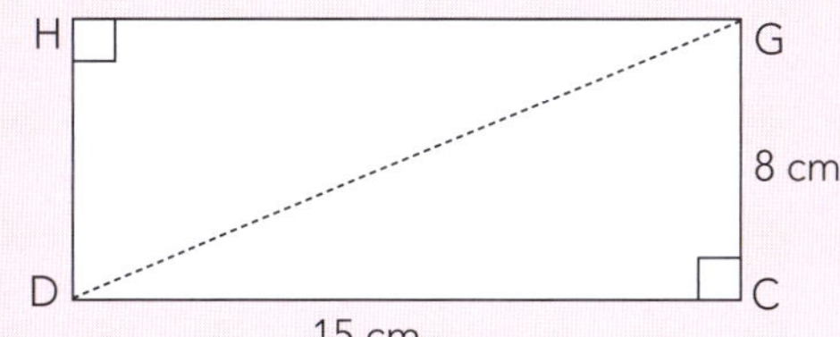

$DG^2 = 15^2 + 8^2$ (Pythagoras)
$\therefore DG = 17$ cm

2 Calculate the length of DF.
DF lies on the triangle DFG.

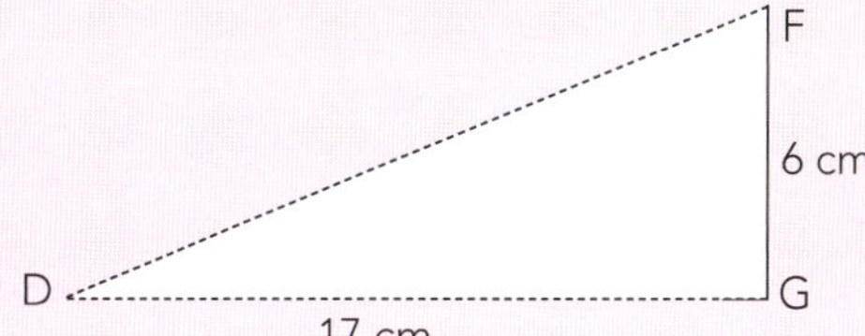

$DF^2 = 17^2 + 6^2$ (Pythagoras)
$\therefore DF = 18.03$ cm

3 Calculate the angle between the line DF and the plane DHGC.
∠FDG lies on triangle DFG.

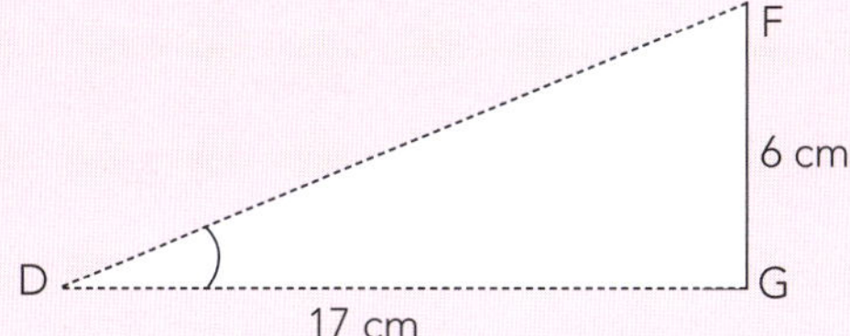

$\tan FDG = \frac{6}{17}$

$\therefore \angle FDG = 19.4°$

4 Calculate the angle between the plane HFD and the plane DHGC.

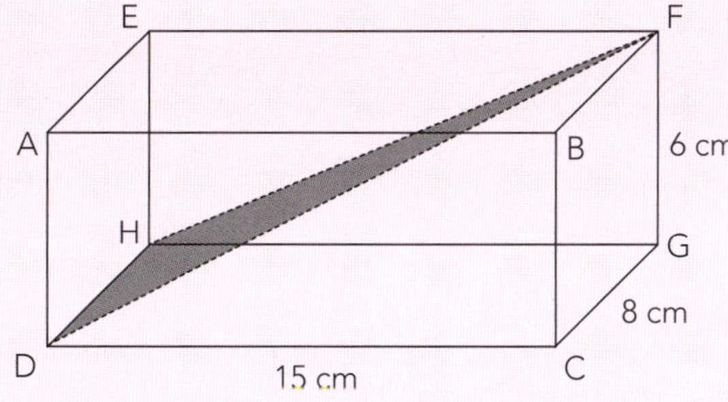

∠FHG lies in the triangle FGH.

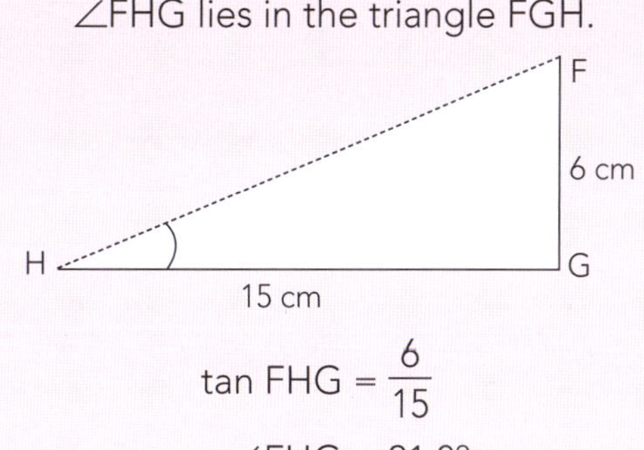

$\tan FHG = \frac{6}{15}$

$\therefore \angle FHG = 21.8°$

ISBN: 9780170370394

Calculate the missing angles and sides for the following.

1 This figure shows a cuboid (box).

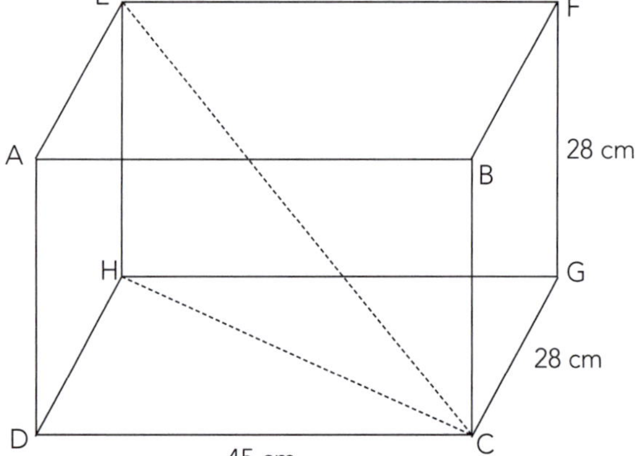

a Calculate the length of HC.

b Calculate the length of EC.

c Calculate the angle between the line EC and the plane HGCD.

d Calculate the angle between the plane ECH and plane EFGH.

e Calculate the angle between the plane EHC and plane AEHD.

 ISBN: 9780170370394

2 A 5.5 m pole (TA) on a roof is being used to support two 8 m long strings of Christmas lights (TB and TD). The bottom ends of the strings of lights (BD) are 8 m apart, and the same distance from A. A third string of lights will be placed from T to a point halfway between B and D (TC).

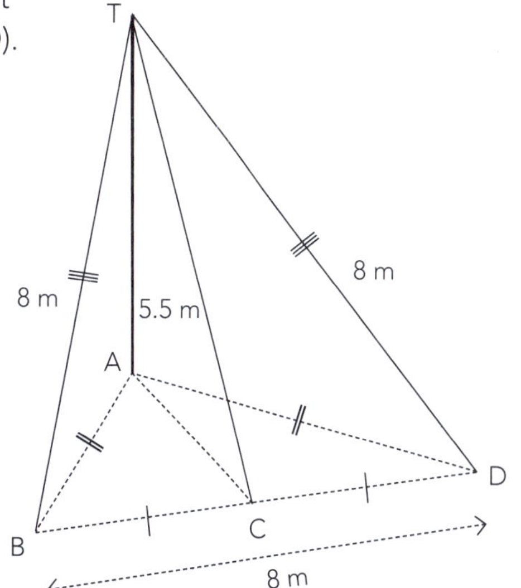

a Calculate the distance between the base of the pole (A) and the end of the 8 m string (D).

____

____

____

b Calculate the angle between the 8 m string of lights (TD) and the roof.

____

____

____

c Calculate the distance between A to the bottom of the middle string of lights.

____

____

____

d How long is TC, the middle string of lights?

____

____

____

e Calculate the angle between TC and the roof.

____

____

____

ISBN: 9780170370394

3 A 6 m-high pyramid is built on a square base with sides that are 6 m long.
The top (T) of the pyramid is 6 m above the centre of the square base.
Some answers will need several steps of working.

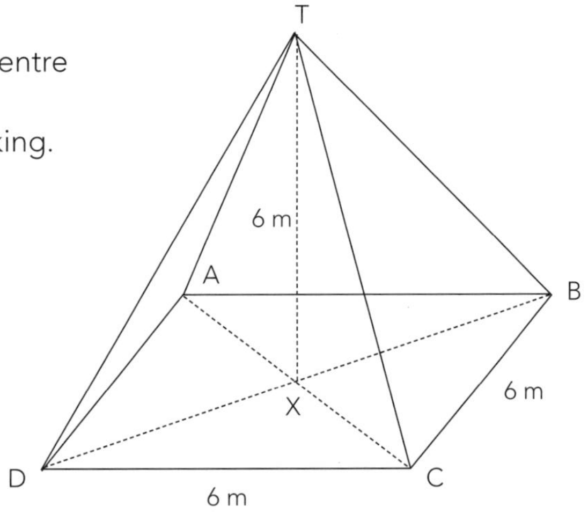

a Calculate the length of the edge TB.

b Calculate the angle between the edge (TB) and the base.

c Calculate angle between the face TCB and the base of the pyramid.

d A second identical pyramid is built next to the original one. Calculate the size of angle TCS.

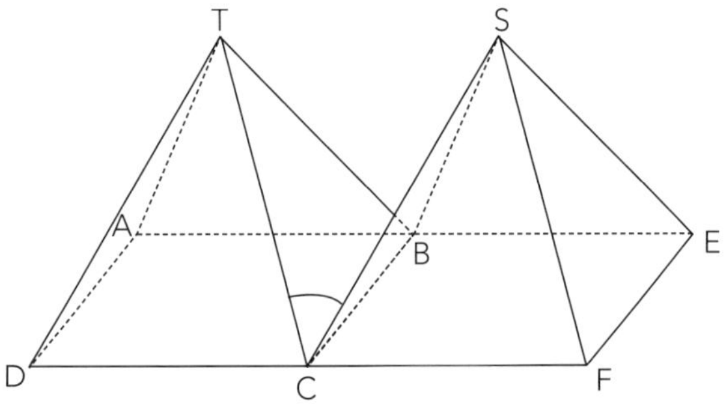

 ISBN: 9780170370394

1.6 Geometry • Flash Cards

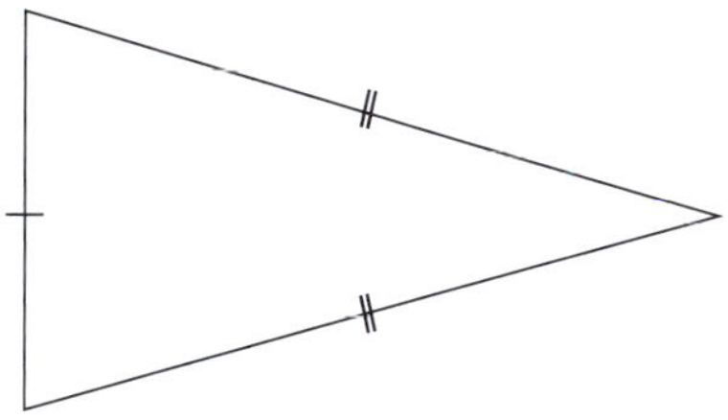

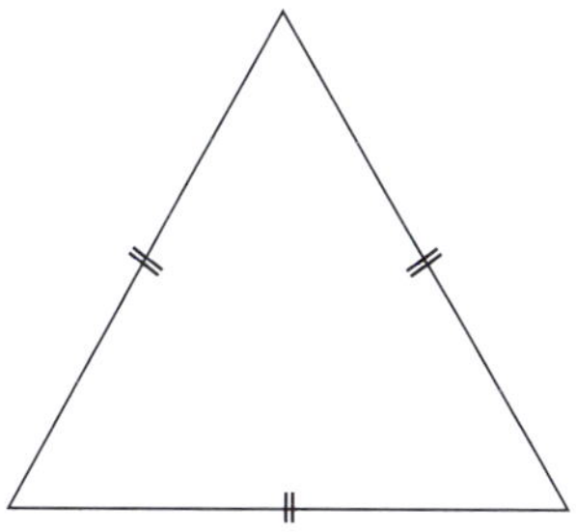

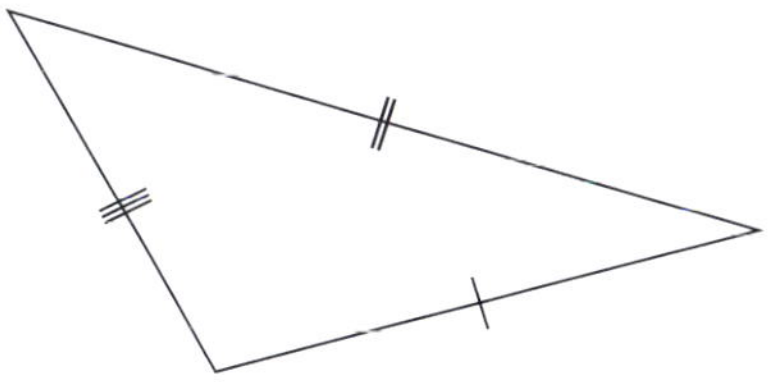

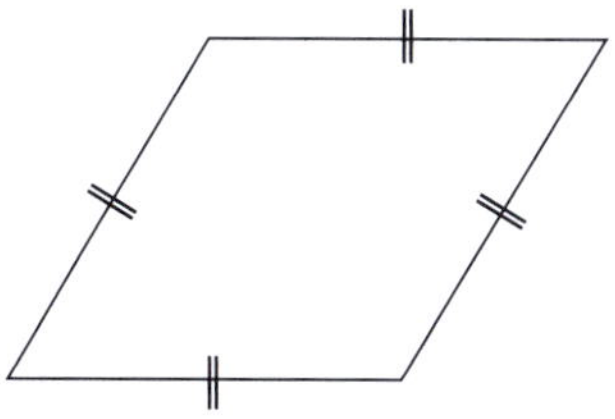

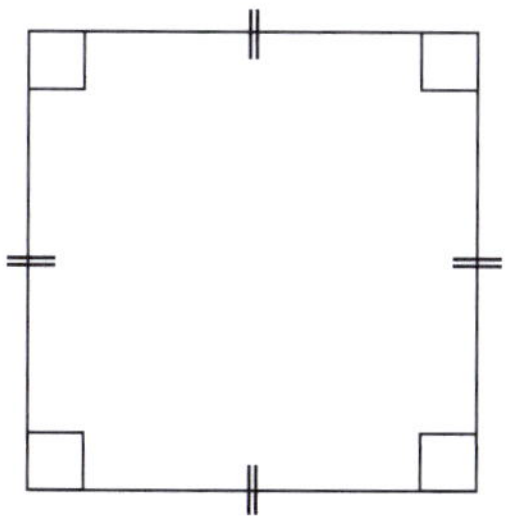

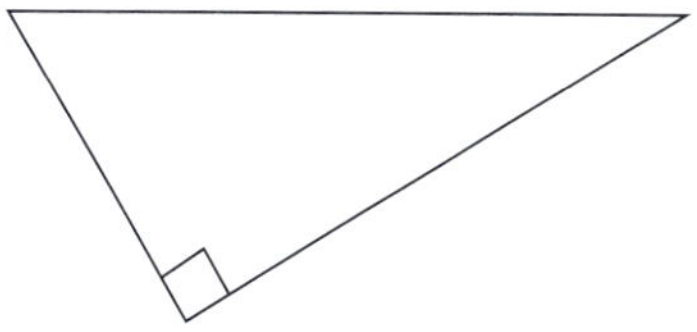

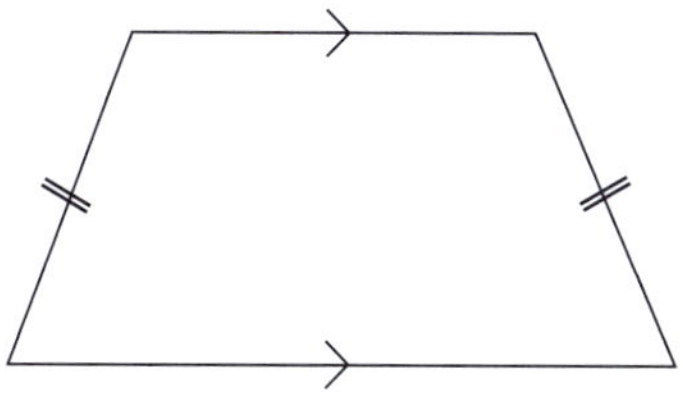

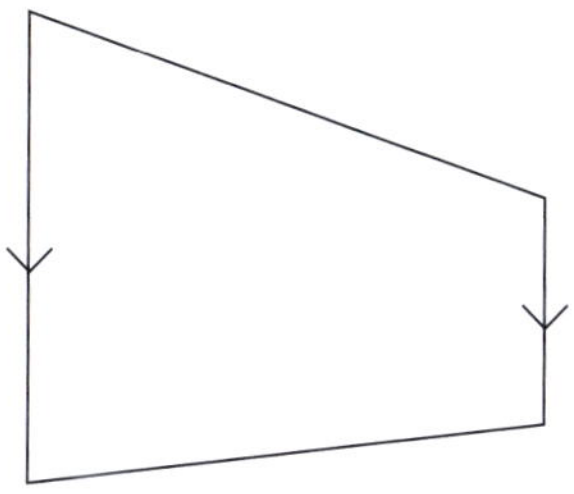

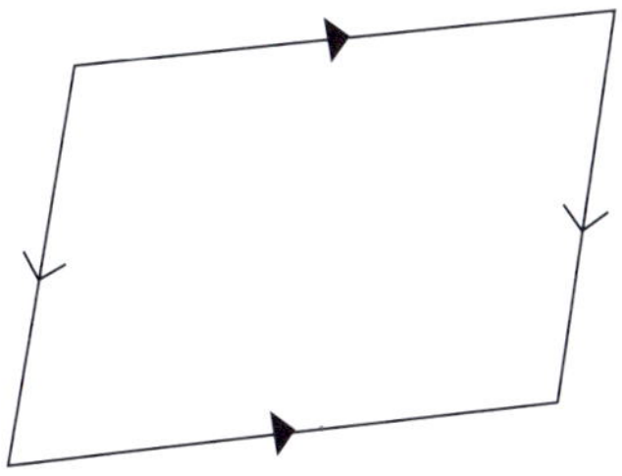

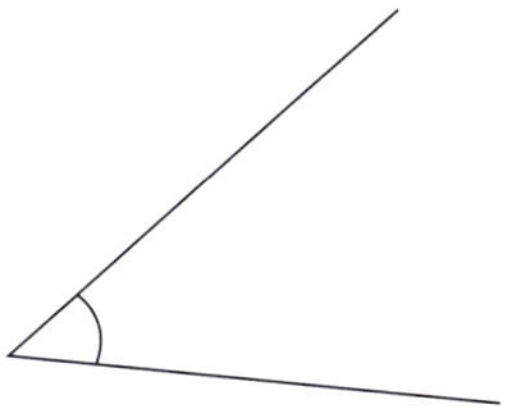

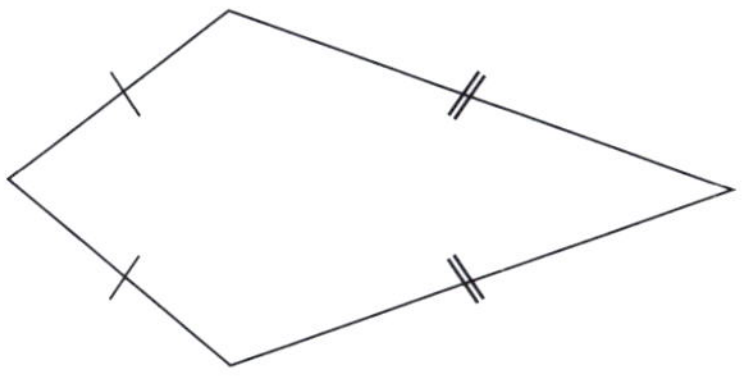

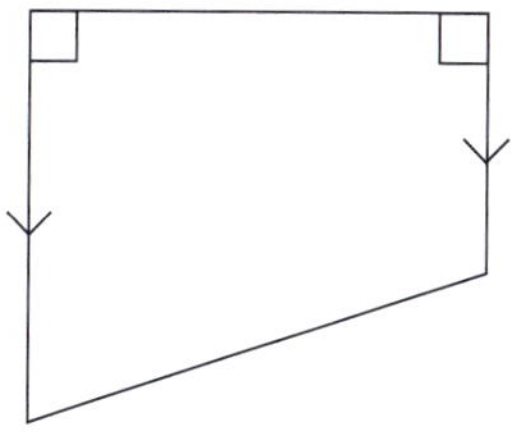

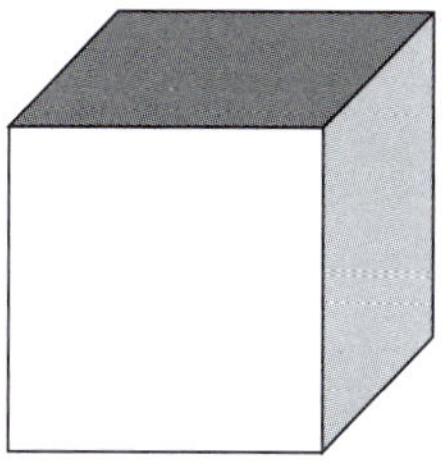

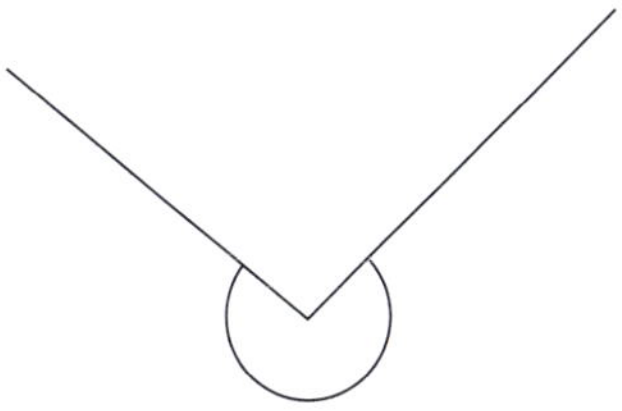

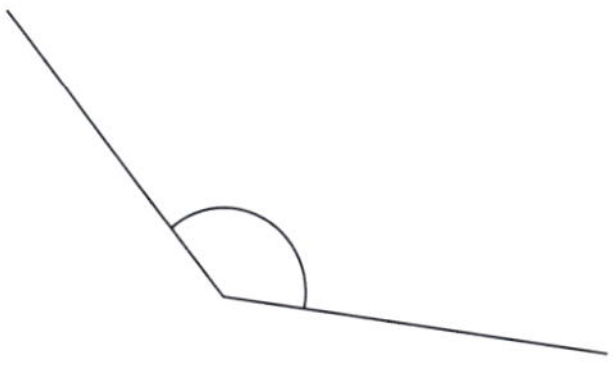

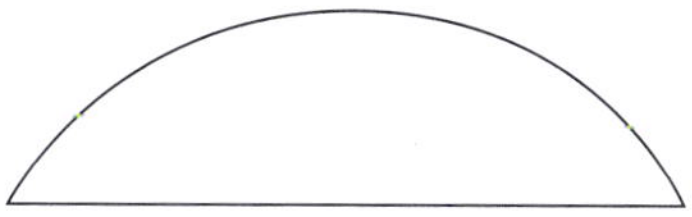

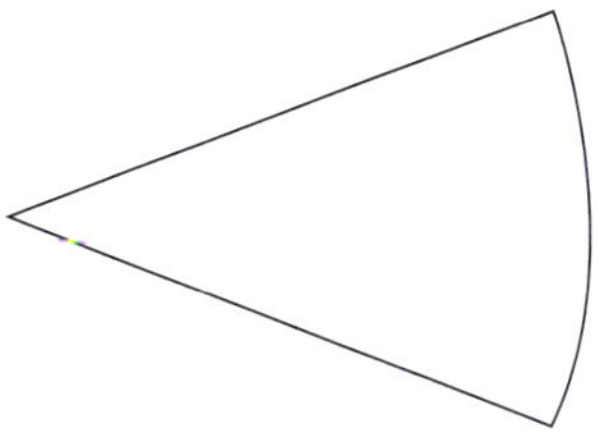

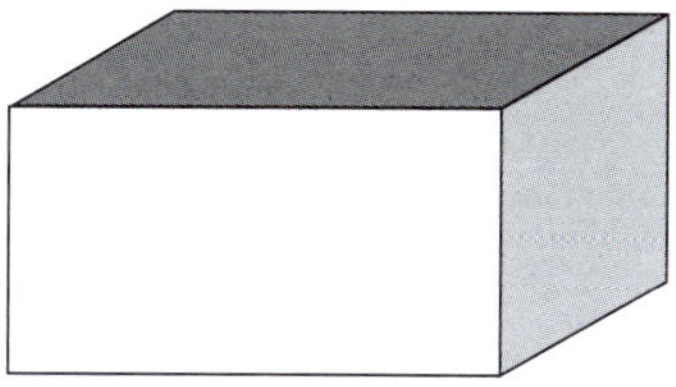

Scalene triangle	Equilateral triangle	Isosceles triangle
Right-angled triangle	Square	Rhombus
Paralellogram	Trapezium	Isosceles trapezium
Right trapezium	Kite	Acute angle
Obtuse angle	Reflex angle	Cube
Cuboid	Sector	Segment

ISBN: 9780170370394

1.6 Geometry • Flash Cards

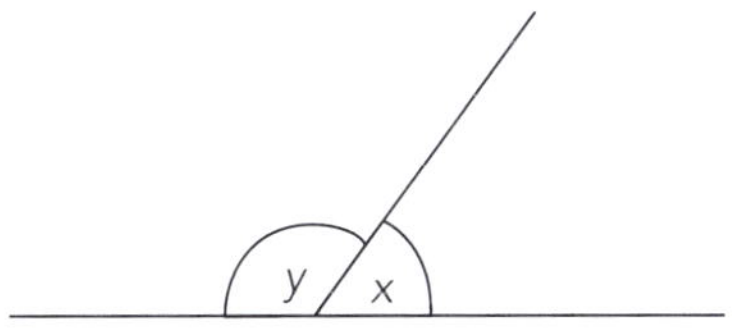

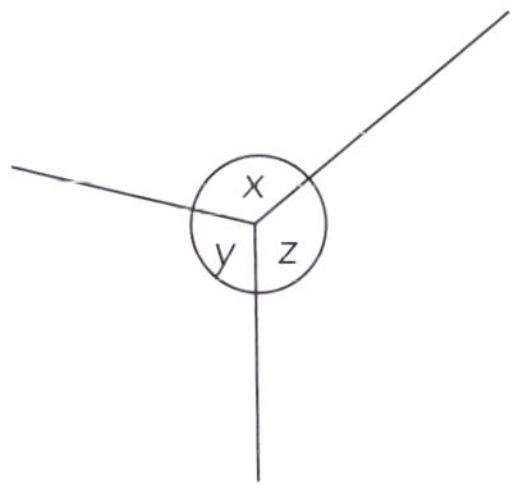

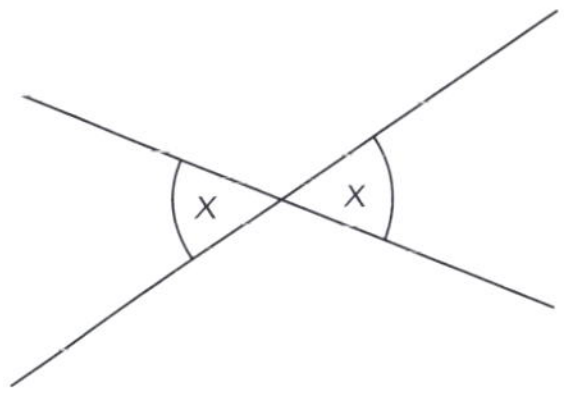

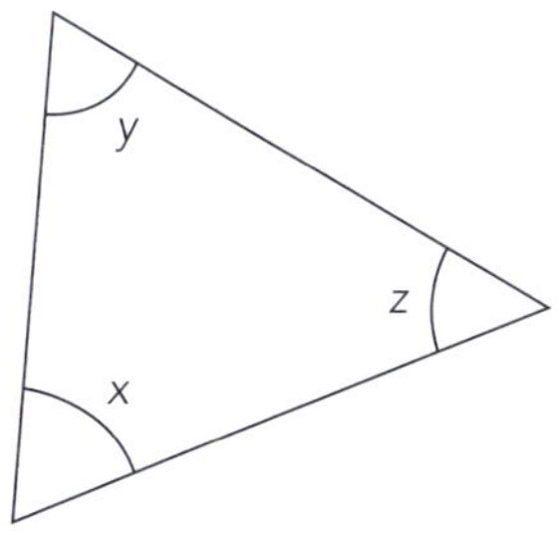

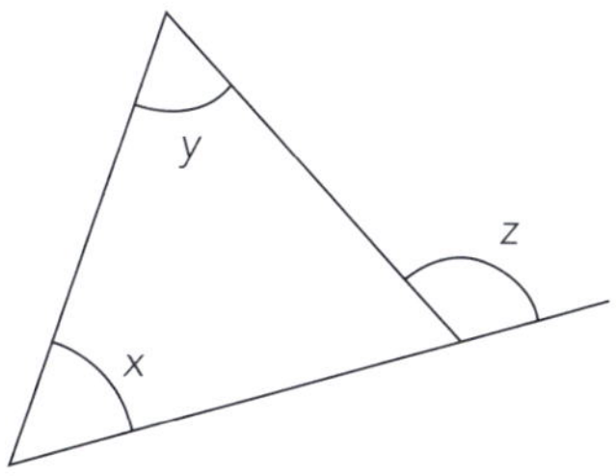

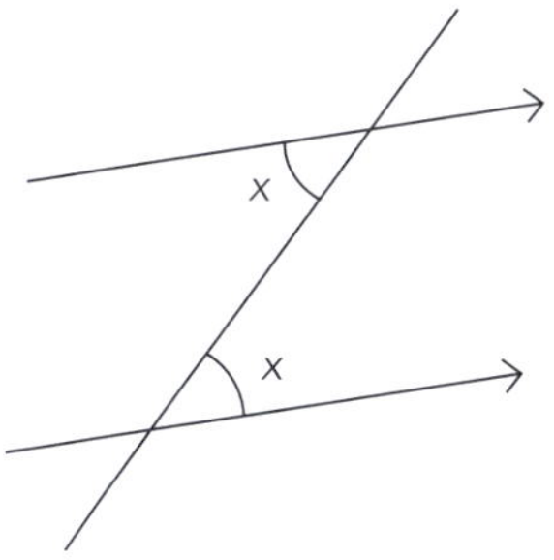

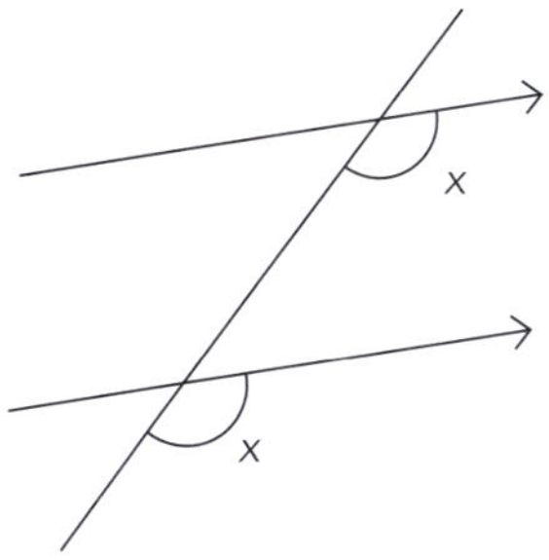

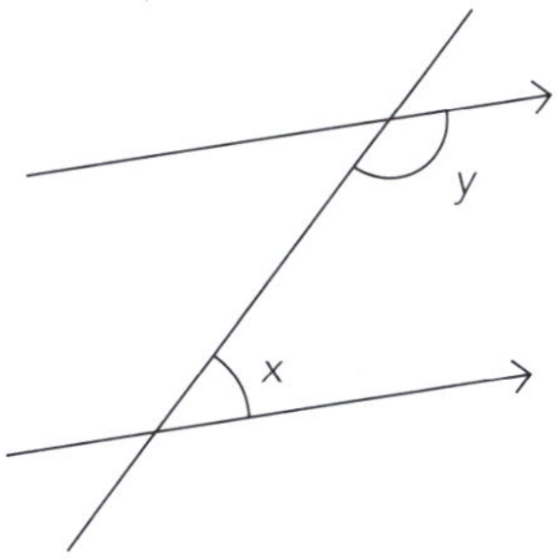

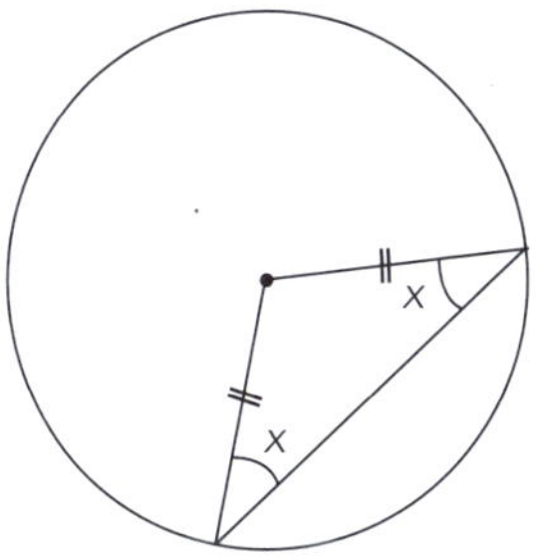

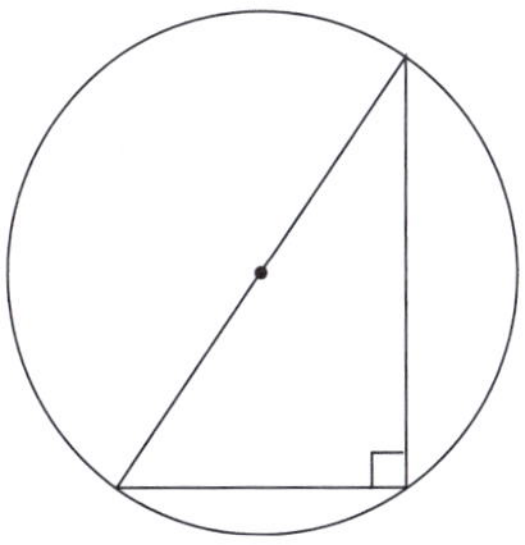

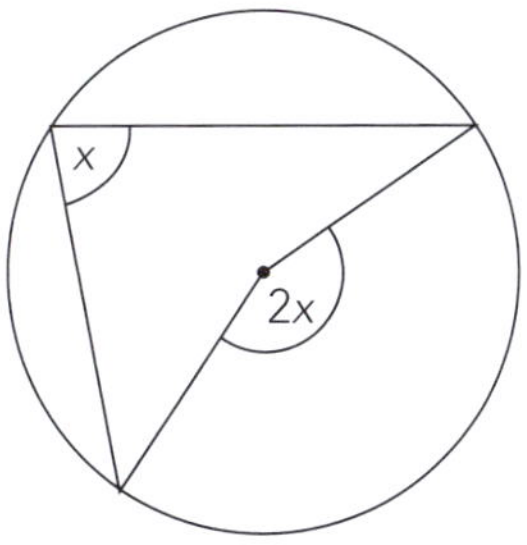

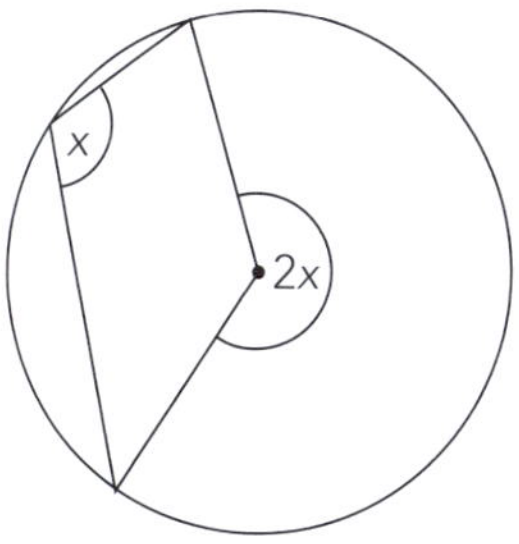

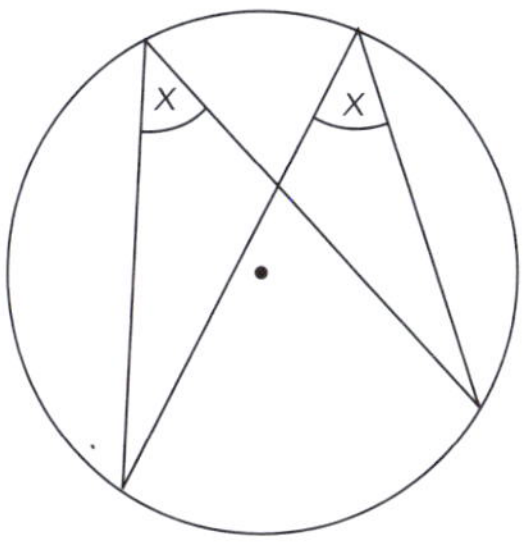

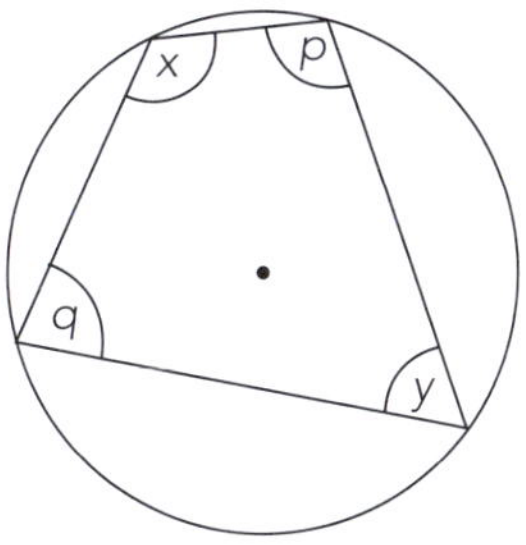

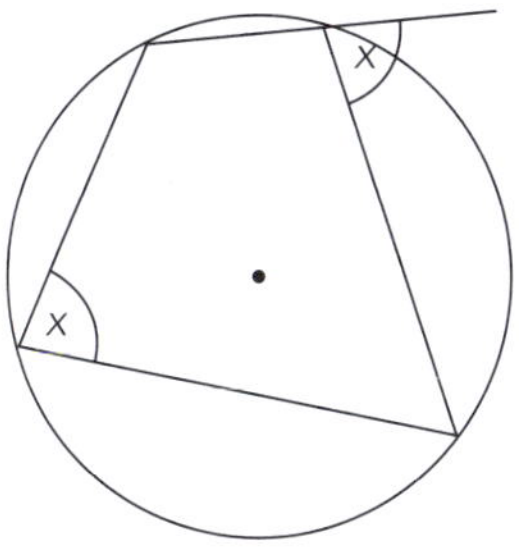

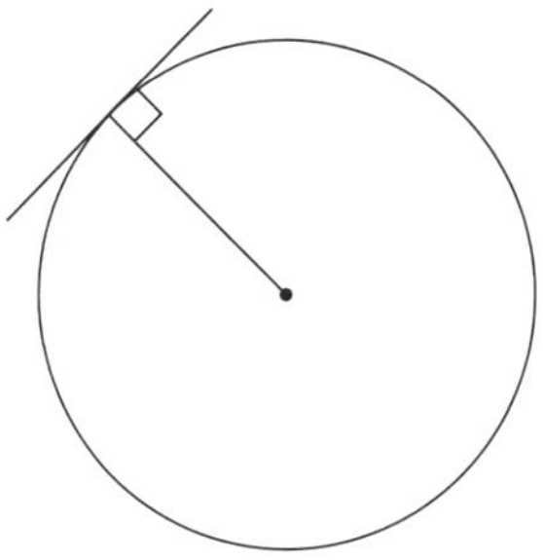

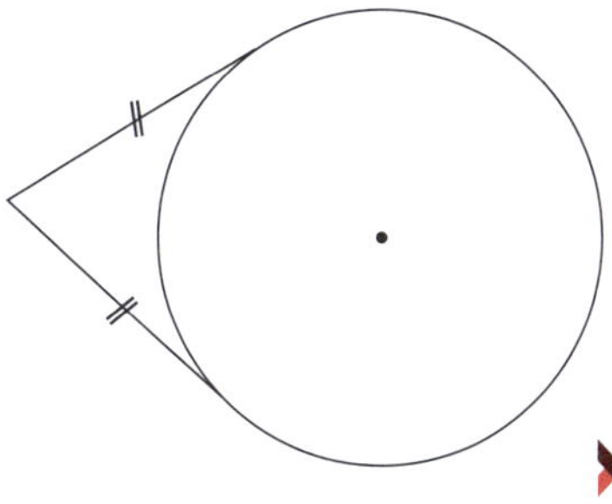

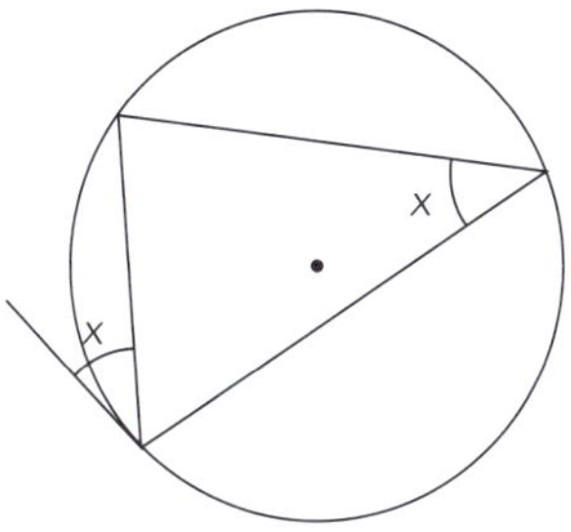

ISBN: 9780170370394

vert opp $\angle$s =	$\angle$s at a point = 360°	$\angle$s on a line = 180°
alt $\angle$s =, // lines	ext $\angle$ of Δ = sum of int $\angle$s	$\angle$s in Δ = 180°
isos Δ, base $\angle$s =	co-int $\angle$s add to 180°, // lines	corr $\angle$s =, // lines
$\angle$ at cent is 2 x $\angle$ at circ	$\angle$ at cent is 2 x $\angle$ at circ	$\angle$ in semicircle = 90°
ext $\angle$ of cyclic quad = the int opp $\angle$	opp $\angle$s of cyclic quad = 180° ($x + y = 180°$ $p + q = 180°$)	$\angle$s on the same arc =
$\angle$ between chord and tan = $\angle$ in the alt seg	= tans	tan $\perp$ rad

ISBN: 9780170370394

Bearings

- Bearings are used in navigation to define direction in a horizontal plane.
- Direction can be given in terms of north, south, east and west.
- Bearings are measured in degrees in a clockwise direction from north.
- All bearings must have three digits.

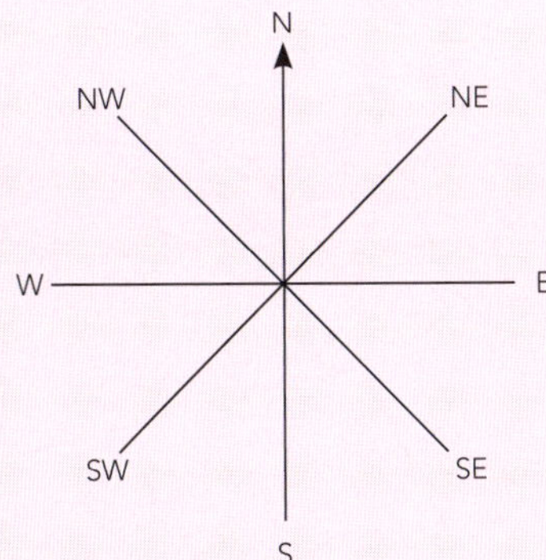

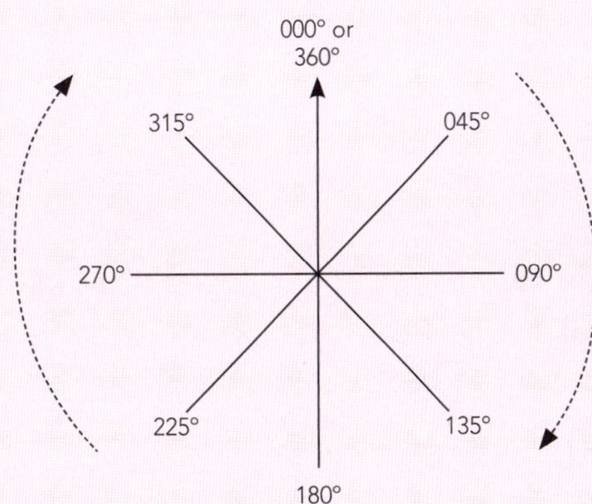

Example one: The first stage (AB) of a cross-country run is 400 m on a bearing of 142°. The second stage (BC) is 550 m on a bearing of 052°.

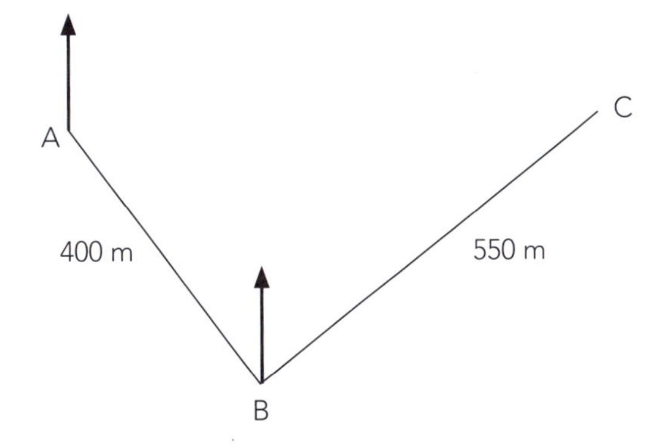

a Calculate the size of angle ABC.

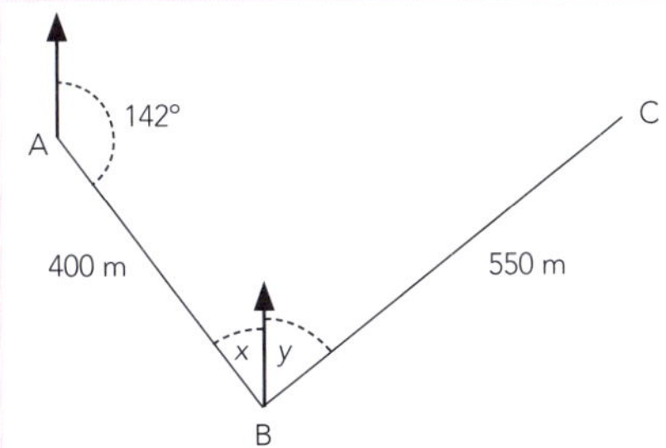

$x = 38°$ (co-int $\angle$s add to 180°, // lines)
$y = 52°$ (bearing given as 052°)

$\therefore$ ABC = 90°

b Calculate the length of the third stage (CA).
The course forms a right-angled triangle.

$$\therefore CA = \sqrt{400^2 + 550^2} = 680.1 \text{ m}$$

ISBN: 9780170370394

c Calculate angle CAB.

$$\angle CAB = \tan^{-1}\frac{550}{400} = 54.0°$$

d Calculate the bearing of the final stage of the course (CA).

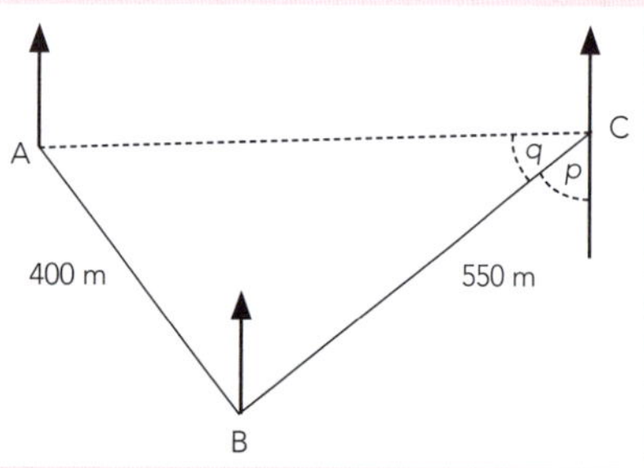

$p = 52°$ (alt $\angle$s =, // lines)
$q = 36°$ ($\angle$s in Δ = 180°)

$\therefore$ Bearing $= 180° + 52° + 36°$
$= 268°$

Example two: Hannah and Brent both started at A. They both walked 150 m, but Hannah walked on a bearing of 208° and Brent walked on a bearing of 102°.

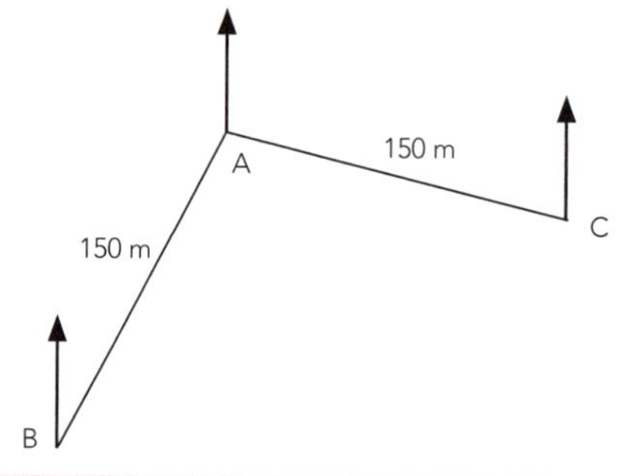

a Calculate the size of angle BAC, and use it to find the size of angle ABC:

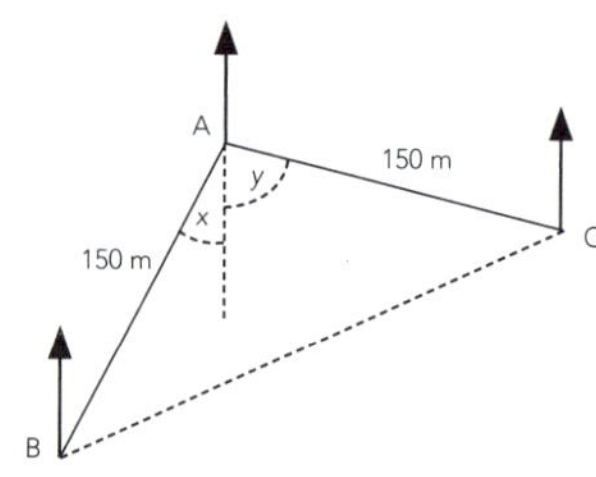

$x = 208° - 180° = 28°$
$y = 180° - 102° = 78°$

$\therefore x + y = 106°$

$\therefore \angle ABC = \frac{180° - 106°}{2} = 37°$ (isos Δ, base $\angle$s =)

b Calculate the distance between Hannah and Brent.

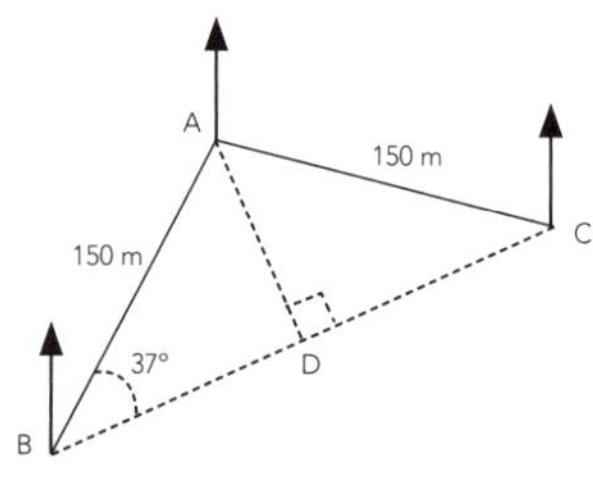

Bisect BC at D to create right-angled triangles ADC and ADB.

$BD = 150 \times \cos 37° = 119.8$ m

$\therefore$ Distance between them = 239.6 m

c Calculate the bearing from Hannah to Brent.

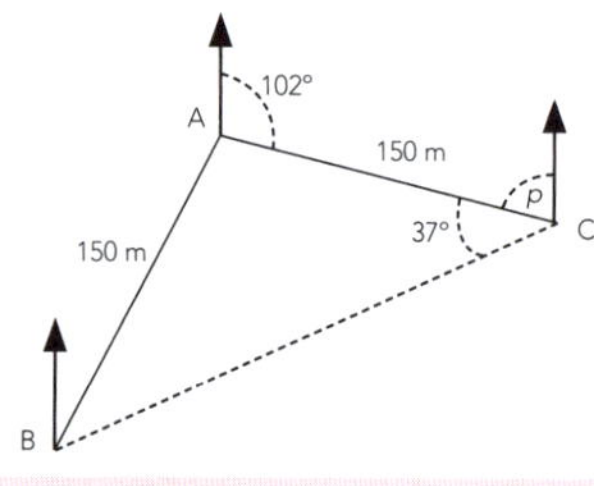

$p = 180° - 102°$
$= 78°$ (co-int $\angle$s add to 180°, // lines)

$\therefore$ Bearing $= 360° - 37° - 78°$
$= 245°$

ISBN: 9780170370394

Answer the following questions. You have been given two copies of each diagram for your working. Show your reasoning for each answer.

1 A yacht race starts from A. The first leg (AB) is 16 nautical miles on a bearing of 040°. The second (BC) is 12 nautical miles on a bearing of 130°. The third leg takes the boats back to the starting point.

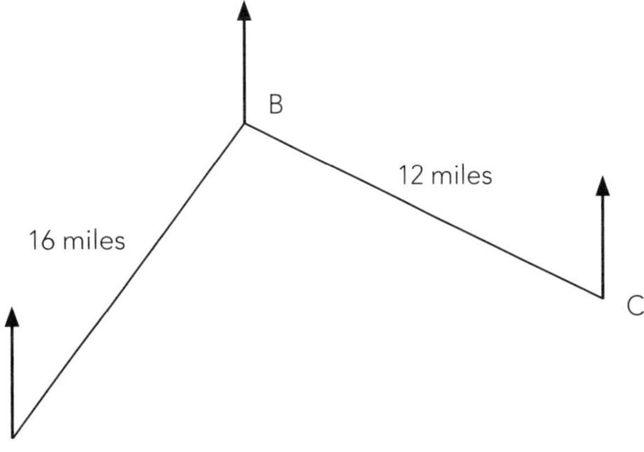

a Calculate the size of the angle ABC.

____

____

b What is the total length of the course?

____

____

c Calculate the angle BCA.

____

____

d Calculate the bearing of the third leg (CA).

____

____

2 A shorter yacht race starts at A, and the first leg is 2.5 nautical miles on a bearing of 342°. The second leg is 3.2 nautical miles on a bearing of 072°. Again, the third leg takes the boats back to the starting point.

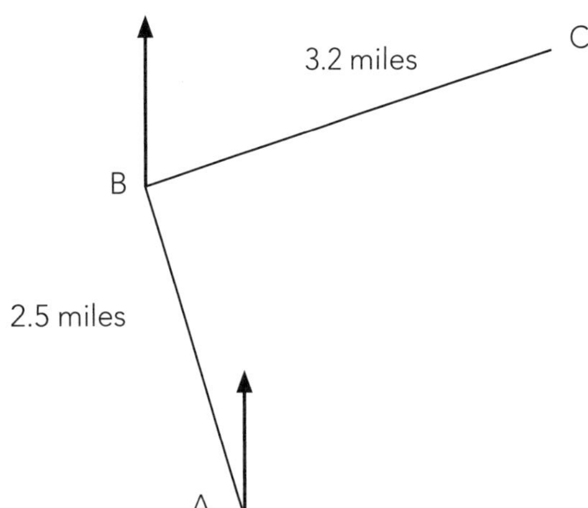

a Calculate the size of the angle ABC.

____

____

b What is the total length of the course?

____

____

c Calculate the angle BCA.

____

____

d Calculate the bearing of the third leg (CA).

____

____

ISBN: 9780170370394

3 The first stage (AB) of a cross-country running course is 210 m on a bearing of 298°. The second (BC) is 125 m on a bearing of 208°. The third stage is CA.

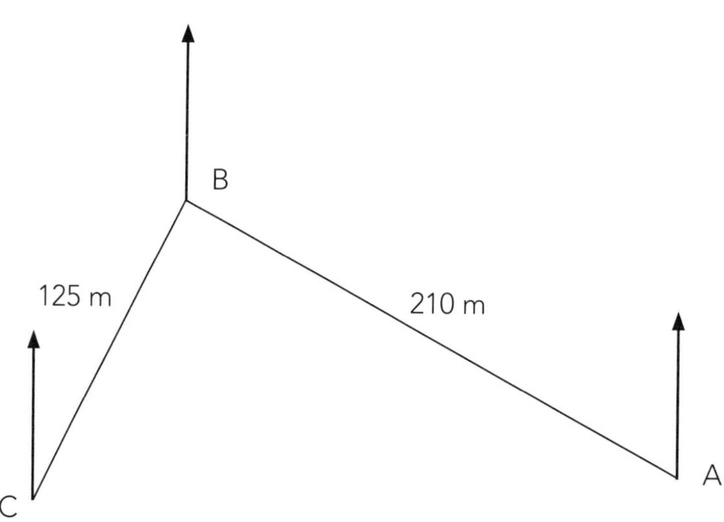

a What is the total length of the course?

____

____

____

____

b Calculate the bearing of the third leg (CA).

____

____

____

____

4 The first stage (QP) of a cross-country running course is 500 m on a bearing of 143°. The second (PR) is 750 m on a bearing of 233°. The third stage is from R to Q.

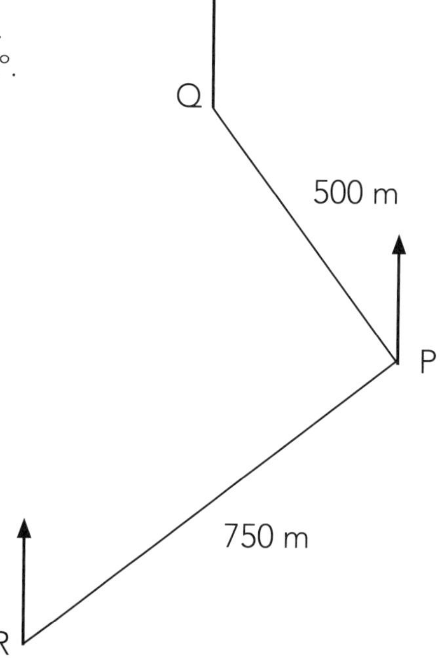

a What is the total length of the course?

____

____

____

____

b Calculate the bearing of the third leg (RQ).

____

____

____

____

ISBN: 9780170370394

5 The first stage (AB) of a sailing course is 1000 m on a bearing of 115°.
The second stage (BC) is the same distance but on a bearing of 353°.

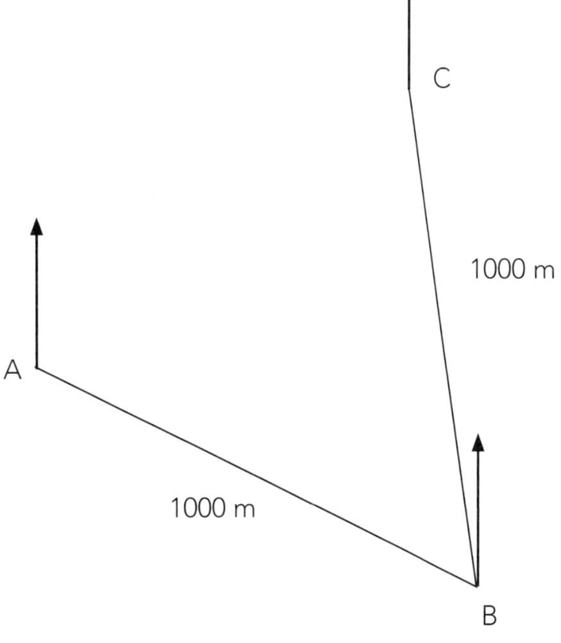

a Calculate the size of the angle ABC.

b What is the total length of the course?

c Calculate the angle BCA.

d Calculate the bearing of the third leg (CA).

6 A cross-country course has the shape of a regular pentagon ABCDE.
The first stage (AB) is 300 m on a bearing of 035°.

a Calculate the bearings for the remaining stages of the course.

b One runner cheats by running directly from A to C. By what distance does he shorten his run?

ISBN: 9780170370394

Angles with circles

Fundamentals

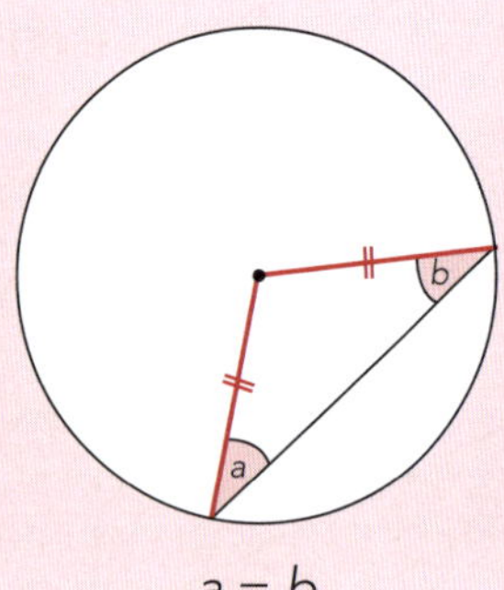

$a = b$

A triangle made of 2 radii is isosceles.

(isos Δ, base ∠s =)

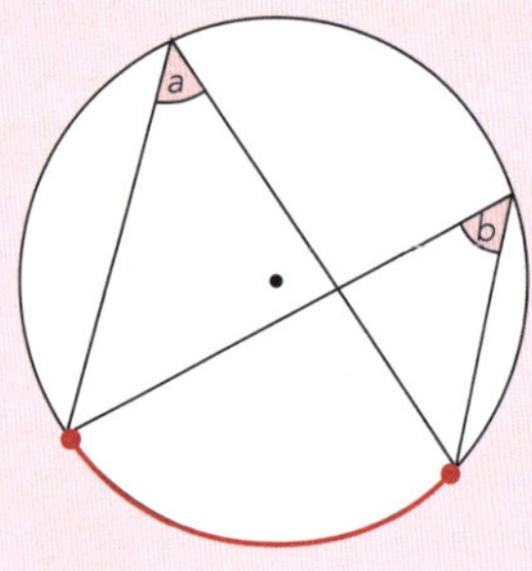

$a = b$

Angles on the same arc are equal.

(∠s on the same arc =)

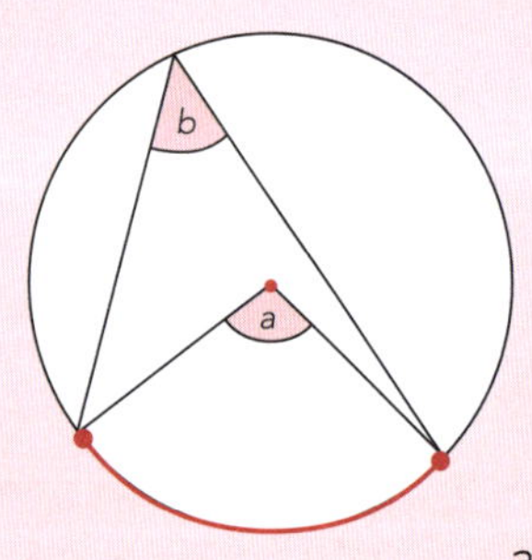

or

The angle at the centre is double the angle at the circumference.

(∠ at cent is 2 x ∠ at circ)

$a = 2b$

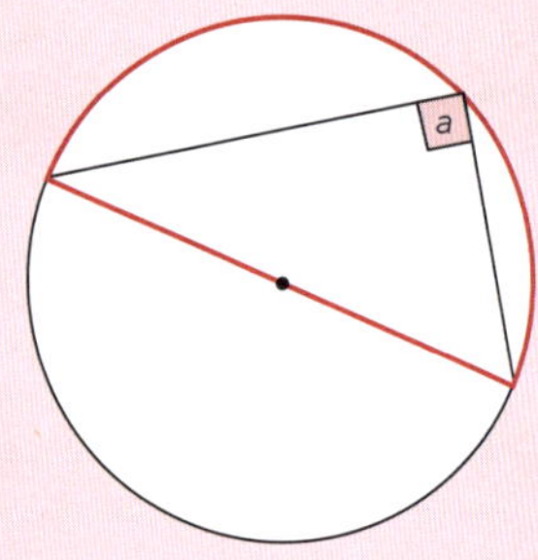

$a = 90°$

The angle in a semicircle is a right angle.

(∠ in semicircle = 90°)

ISBN: 9780170370394

Example one:

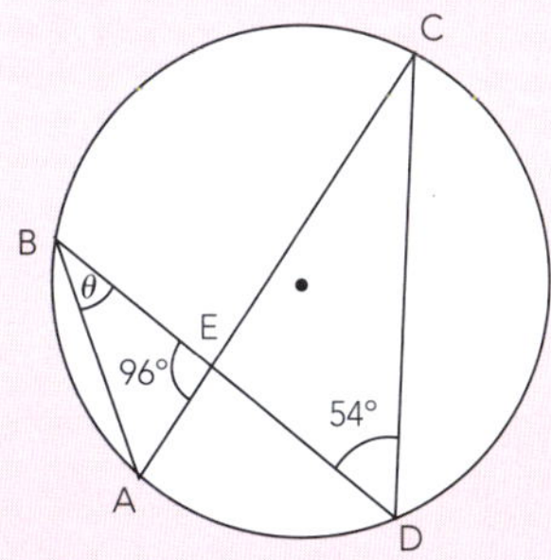

∠CAB = 54° (∠s on the same arc =)

∴ θ = 30° (∠s in Δ = 180°)

Example two:

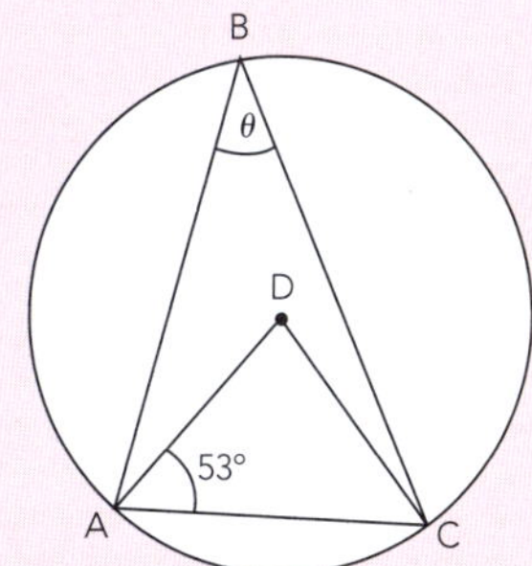

∠DCA = 53° (isos Δ, base ∠s =)

∠ADC = 74° (∠s in Δ = 180°)

∴ θ = 37° (∠ at cent is 2 x ∠ at circ)

Example three:

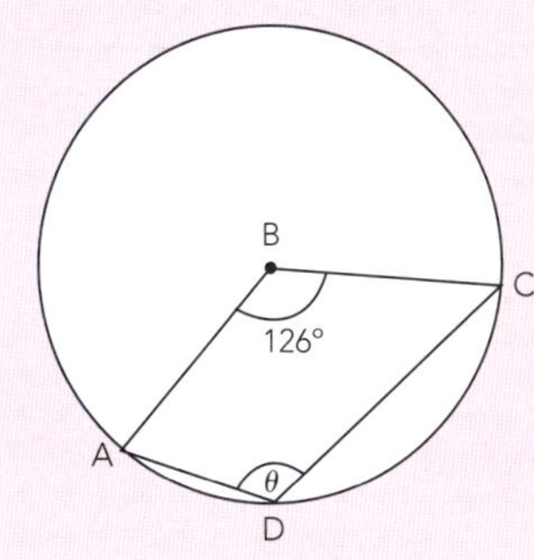

Reflex ∠ABC = 234° (∠s at a point = 360°)

∴ θ = 117° (∠ at cent is 2 x ∠ at circ)

Example four:

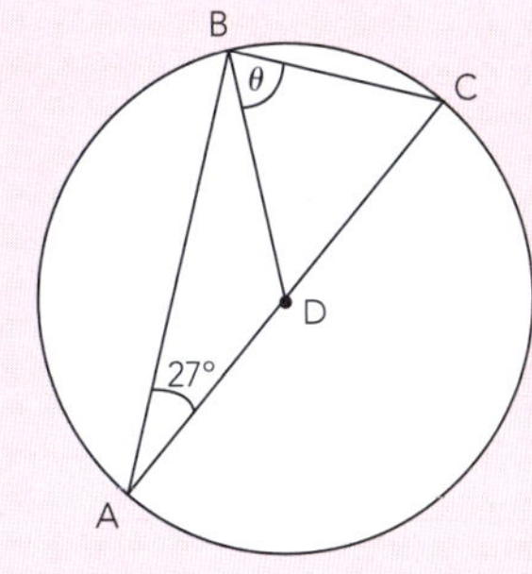

∠BCD = 63° (∠ in semicircle = 90° and
∠s in Δ = 180°)

∴ θ = 63° (isos Δ, base ∠s =)

Calculate the unknown angle(s) and give reasons.

1

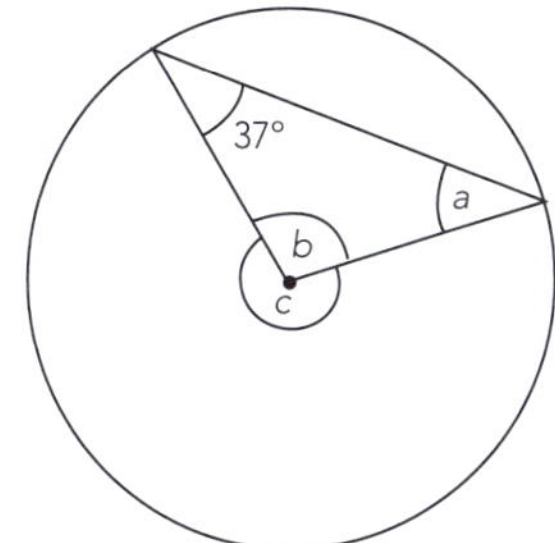

a = ___ ______________________________

b = ___ ______________________________

c = ___ ______________________________

ISBN: 9780170370394

2

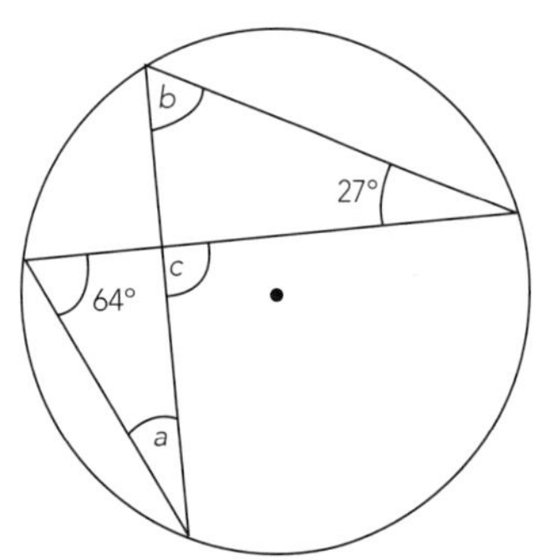

$a =$ ___ ________________

$b =$ ___ ________________

$c =$ ___ ________________

3

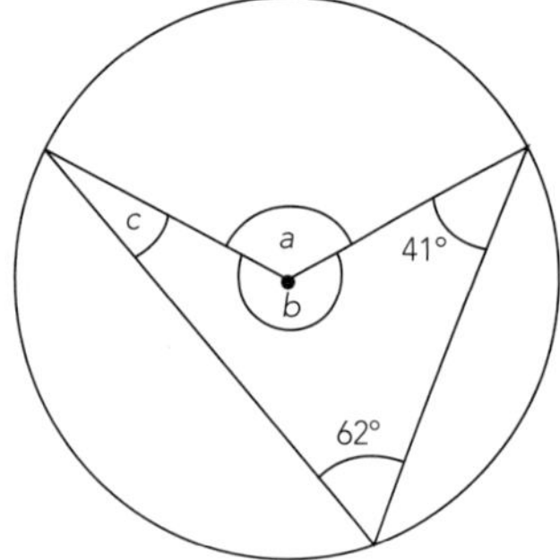

$a =$ ___ ________________

$b =$ ___ ________________

$c =$ ___ ________________

4

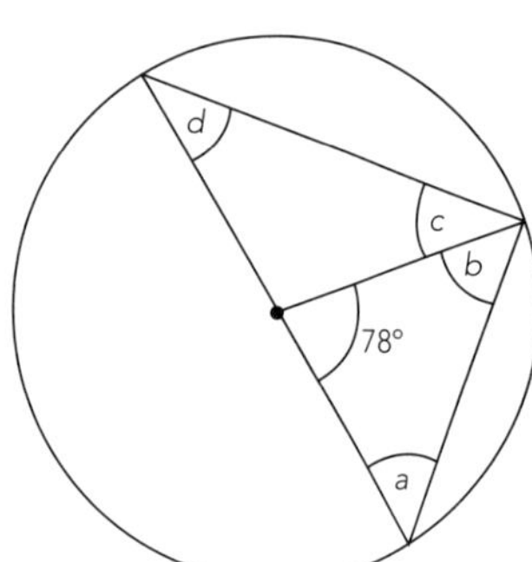

$a =$ ___ ________________

$b =$ ___ ________________

$c =$ ___ ________________

$d =$ ___ ________________

5

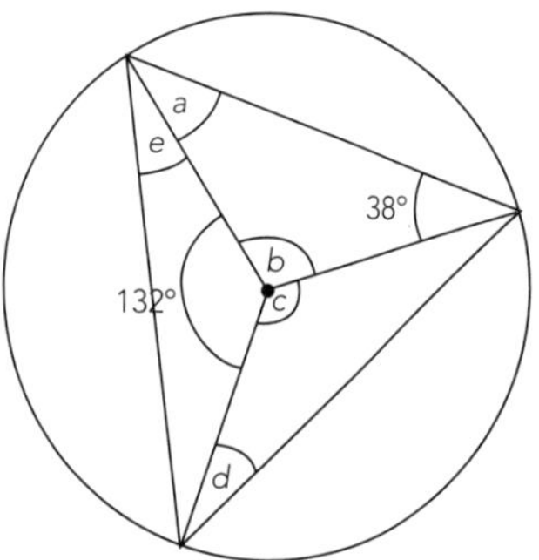

$a =$ ___ ________________

$b =$ ___ ________________

$c =$ ___ ________________

$d =$ ___ ________________

$e =$ ___ ________________

6

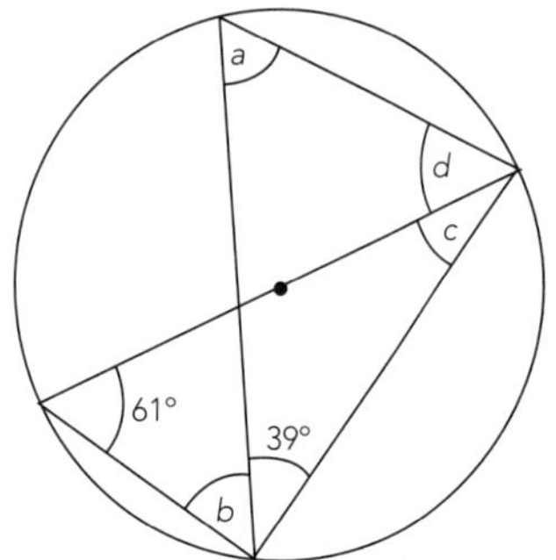

$a =$ ___ ________________

$b =$ ___ ________________

$c =$ ___ ________________

$d =$ ___ ________________

 ISBN: 9780170370394

7

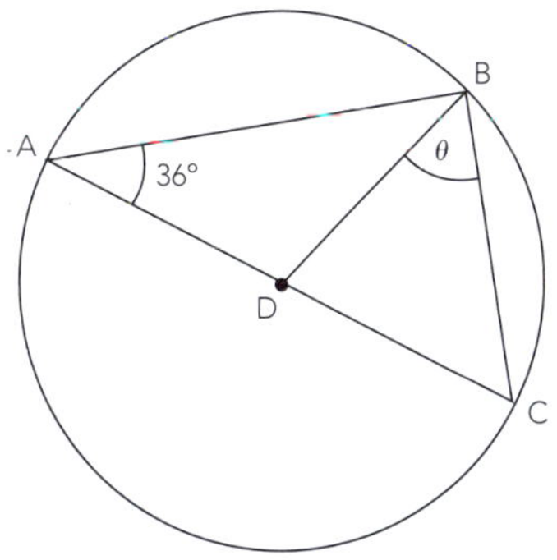

A, B, C, D, 36°, θ

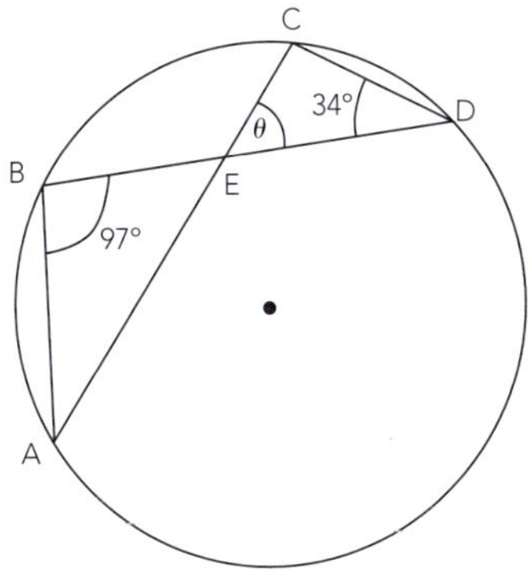

8

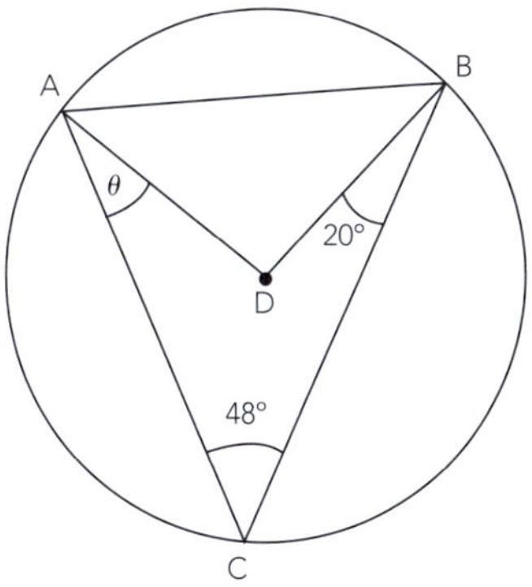

A, B, C, D, θ, 20°, 48°

9

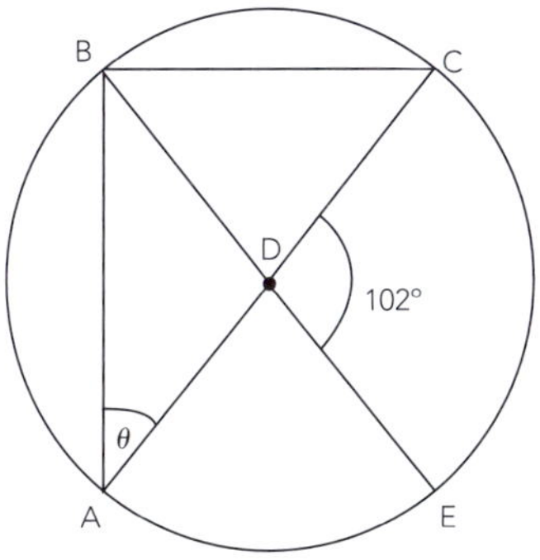

B, C, D, A, E, 102°, θ

10

C, D, B, E, A, 34°, θ, 97°

11

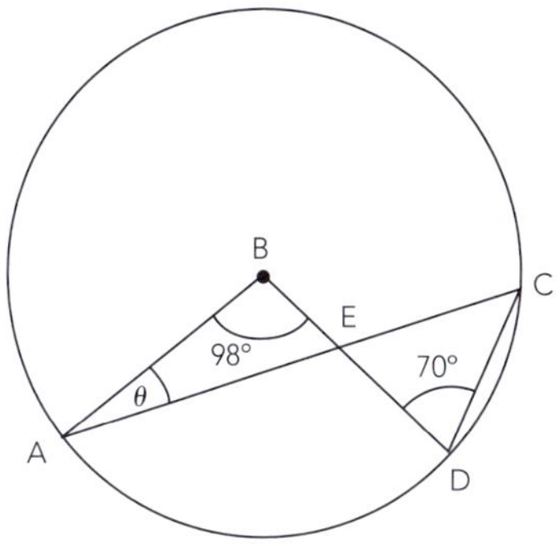

B, C, E, A, D, 98°, 70°, θ

ISBN: 9780170370394

Angles involving tangents

- A tangent is a line that touches a circle in exactly one place.

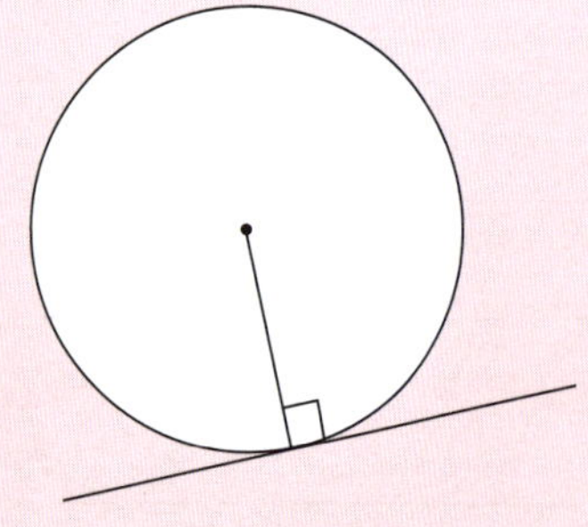

The angle between a tangent and a radius = 90°.

(tan ⊥ rad)

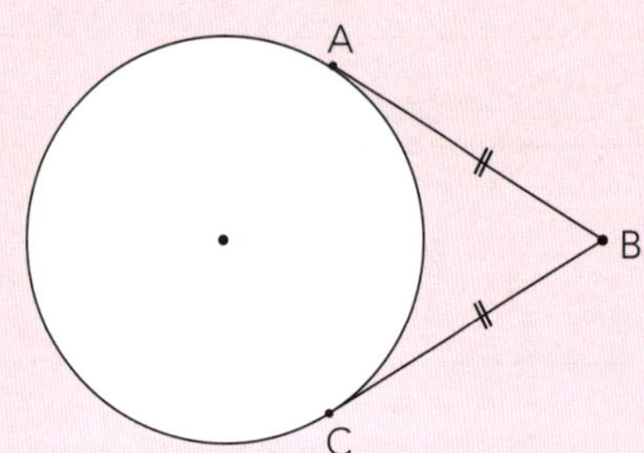

AB = CB

Tangents from a point to a circle are equal.

(= tans)

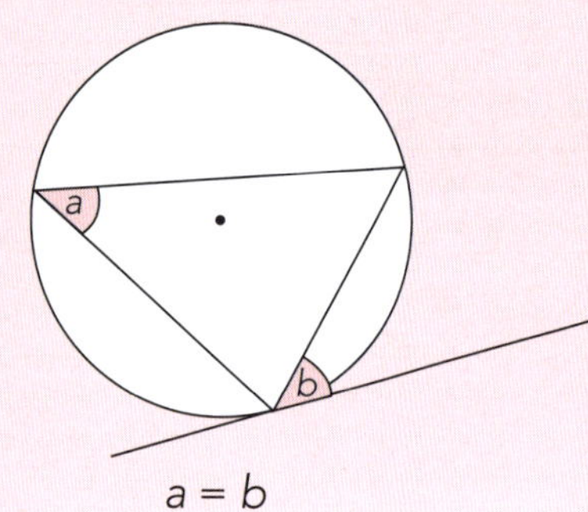

$a = b$

The angle between a chord and a tangent = the angle in the alternate segment.

(∠ between chord and tan = ∠ in the alt seg)

Example one:

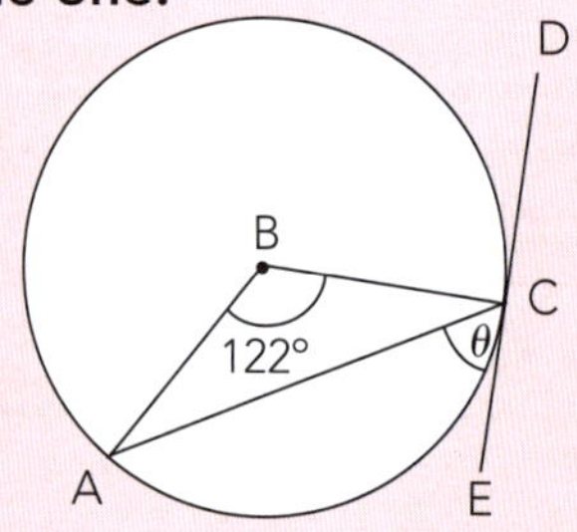

∠BCA = 29° (isos Δ, base ∠s =)

∴ θ = 61° (tan ⊥ rad)

Example two:

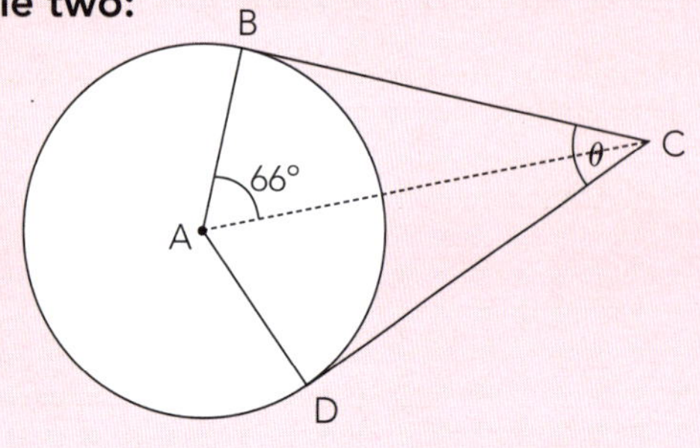

∠BAD = 132° (equal tangents ⟹ ΔABC congruent to ΔADC)

∴ θ = 48° (tan ⊥ rad and ∠s of quad = 360°)

 ISBN: 9780170370394

Example three:

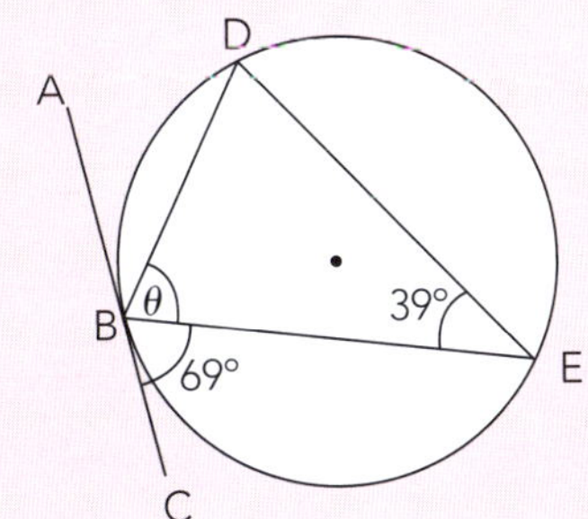

$\angle BDE = 69°$ ($\angle$ between chord and tan = $\angle$ in the alt seg)

$\therefore\ \theta = 72°$ ($\angle$s in $\Delta = 180°$)

Calculate the unknown angle(s) and give reasons.

1

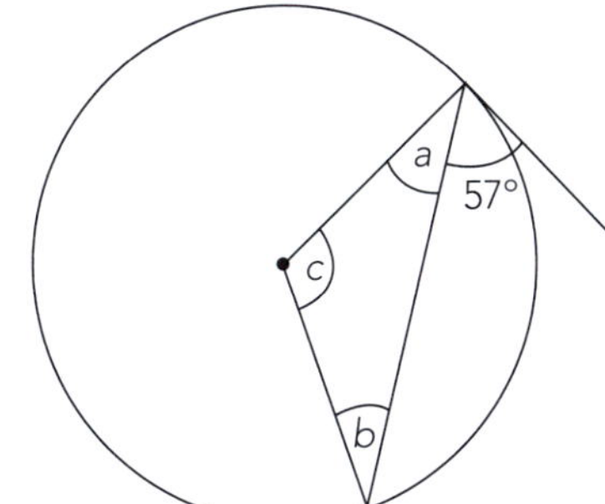

a = ___ ______________________

b = ___ ______________________

c = ___ ______________________

2

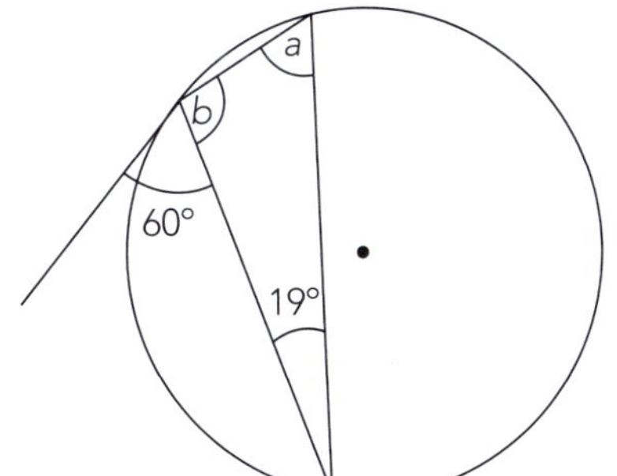

a = ___ ______________________

b = ___ ______________________

3

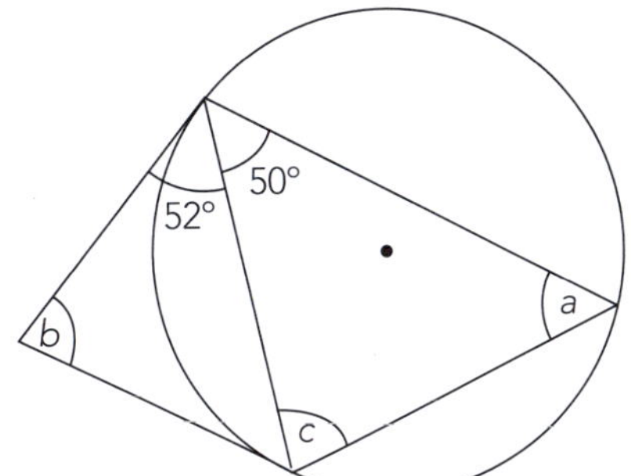

a = ___ ______________________

b = ___ ______________________

c = ___ ______________________

4

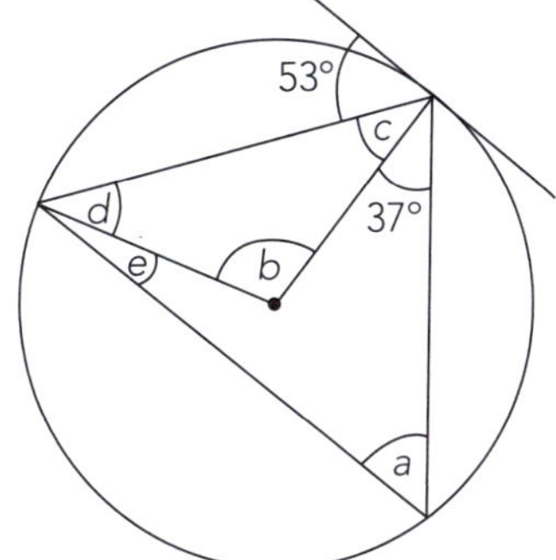

a = ___ ______________________

b = ___ ______________________

c = ___ ______________________

d = ___ ______________________

e = ___ ______________________

ISBN: 9780170370394

5

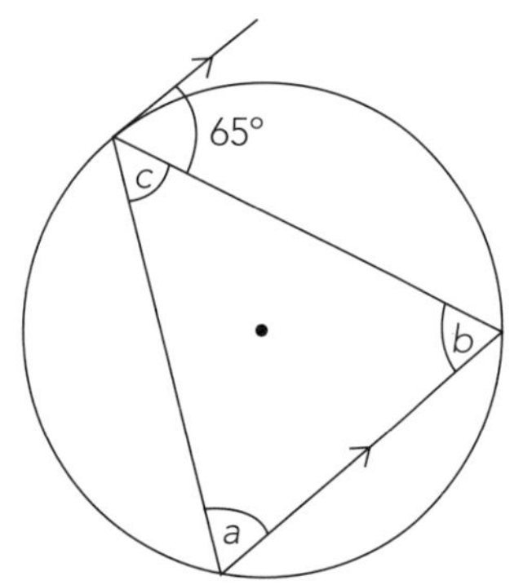

$a =$ ___ ______________________

$b =$ ___ ______________________

$c =$ ___ ______________________

6

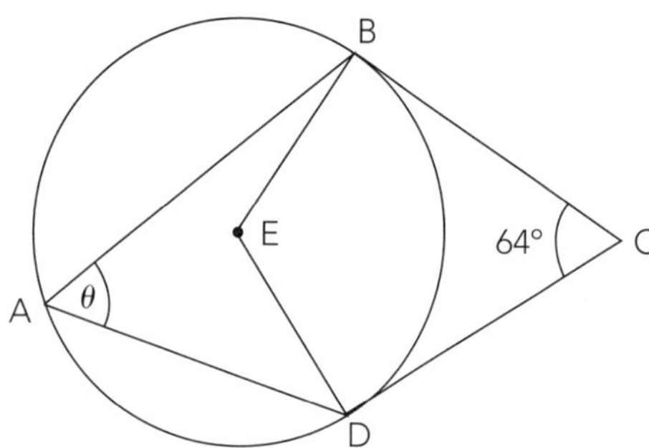

7

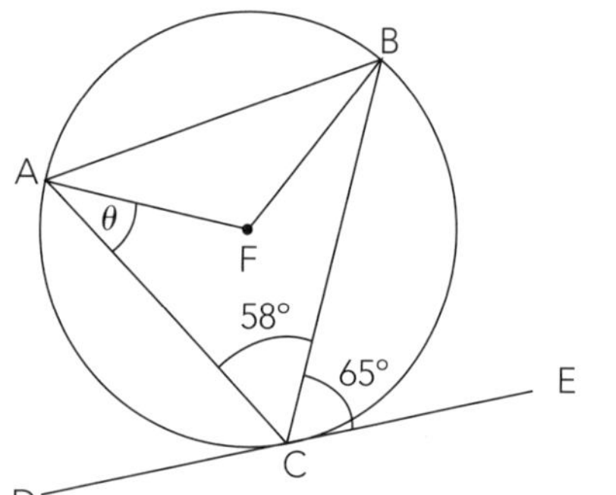

8

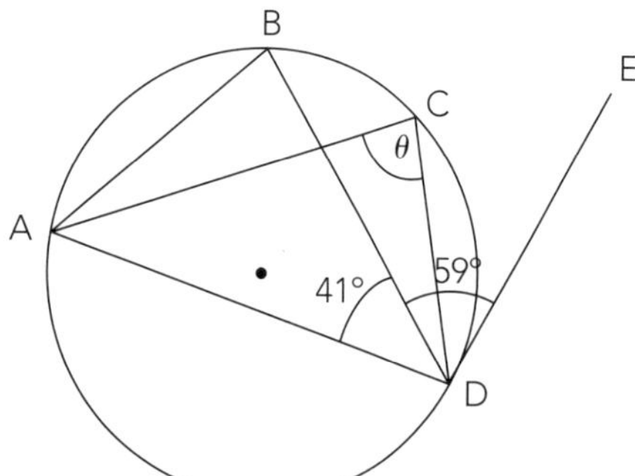

9

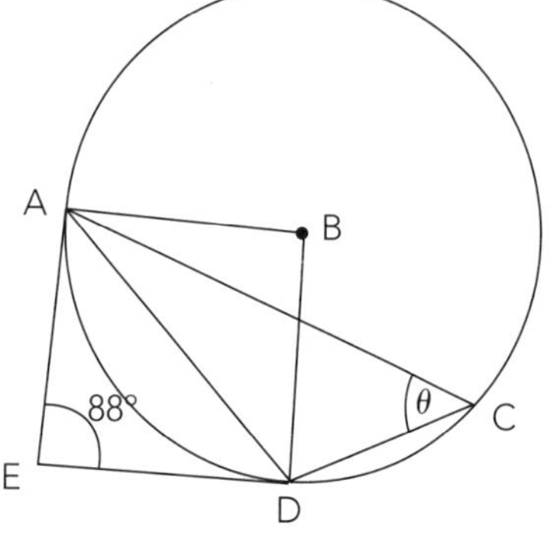

 ISBN: 9780170370394

Angles in cyclic quadrilaterals

- A cyclic quadrilateral is a quadrilateral whose vertices lie on the circumference of a circle.

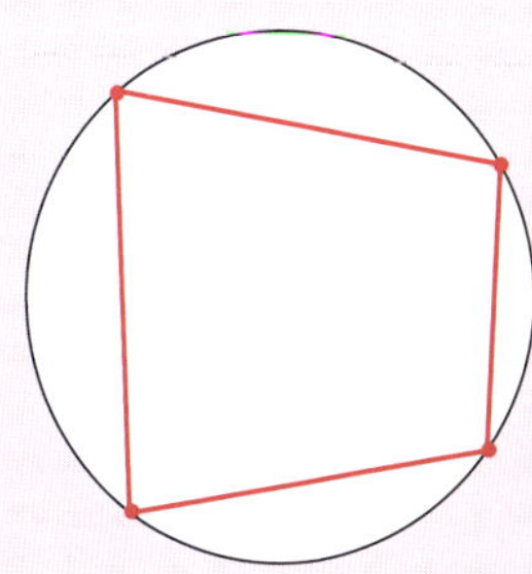

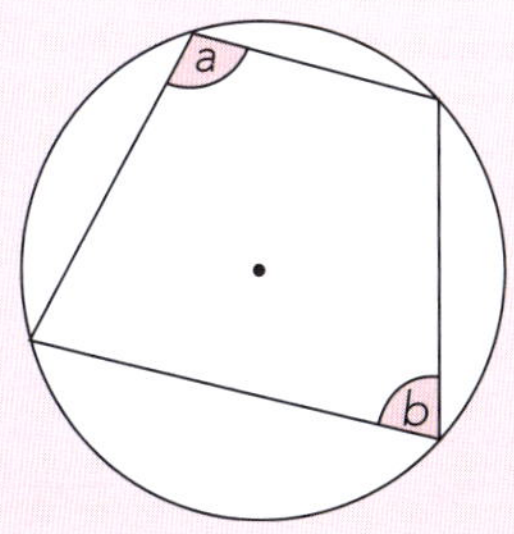

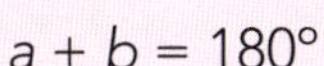
$a + b = 180°$

Opposite angles of a cyclic quadrilateral add to 180°.

(opp ∠s of cyclic quad add to 180°)

Because angles on a line add to 180°, this also means:

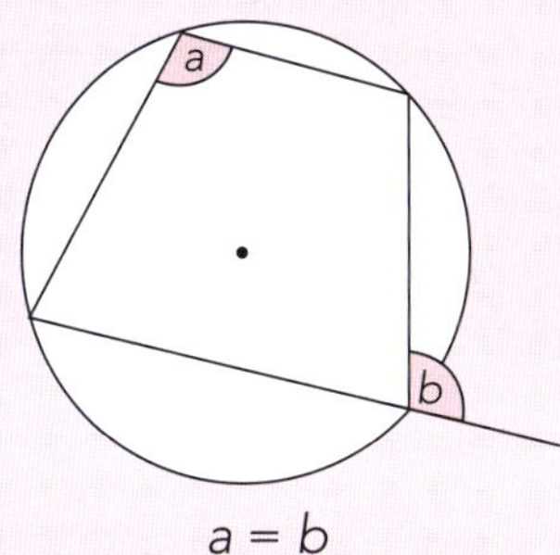

$a = b$

The exterior angle of a cyclic quadrilateral is equal to the internal opposite angle.

(ext ∠ of cyclic quad = the int opp ∠)

Example one:

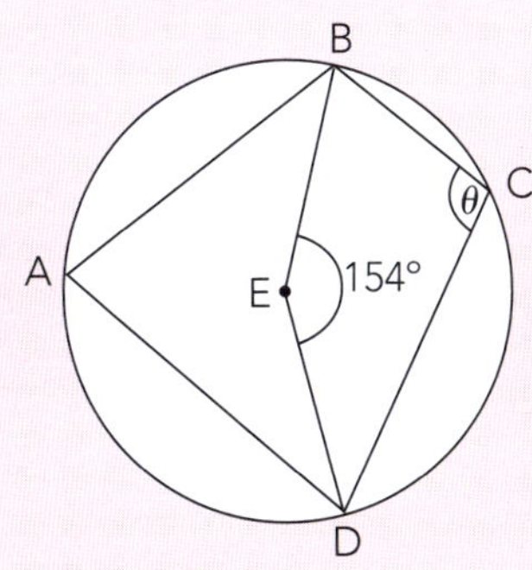

∠BAD = 77° (∠ at cent is 2 x ∠ at circ)

∴ θ = 103° (opp ∠s of cyclic quad add to 180°)

Example two:

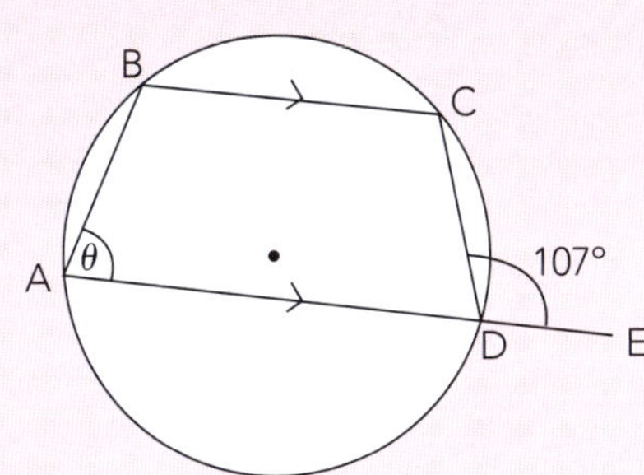

∠ABC = 107° (ext ∠ of cyclic quad = the int opp ∠)

∴ θ = 73° (co-int ∠s add to 180°, // lines)

Calculate the unknown angle(s) and give reasons.

1

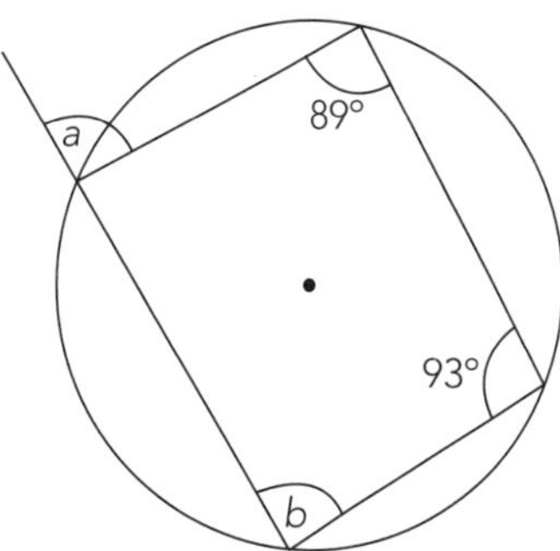

$a =$ ____ ________________________________

$b =$ ____ ________________________________

2

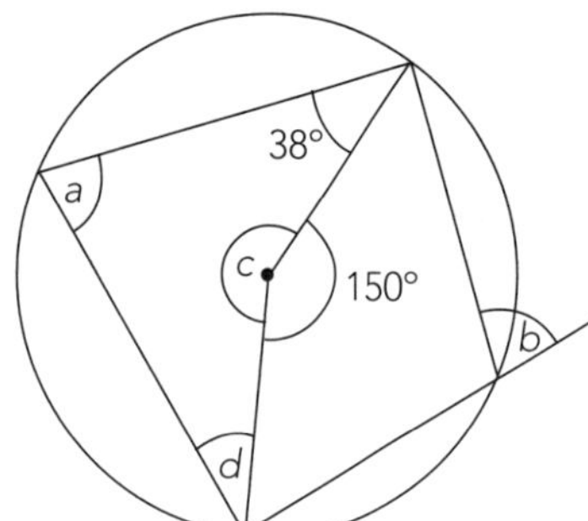

$a =$ ____ ________________________________

$b =$ ____ ________________________________

$c =$ ____ ________________________________

$d =$ ____ ________________________________

3

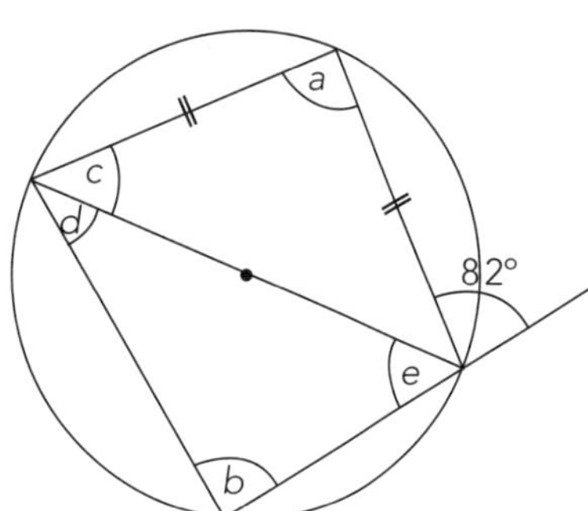

$a =$ ____ ________________________________

$b =$ ____ ________________________________

$c =$ ____ ________________________________

$d =$ ____ ________________________________

$e =$ ____ ________________________________

4

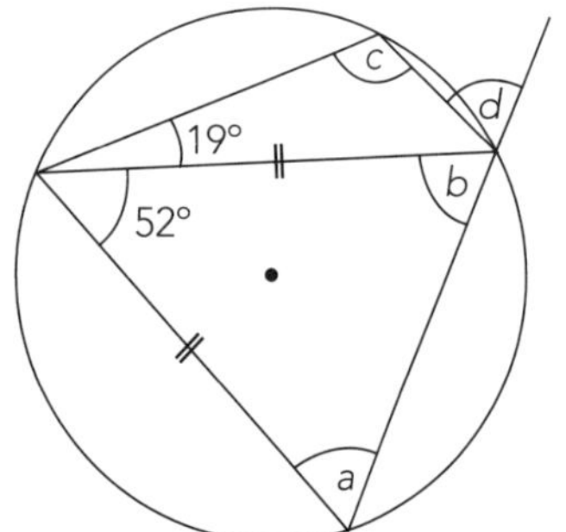

$a =$ ____ ________________________________

$b =$ ____ ________________________________

$c =$ ____ ________________________________

$d =$ ____ ________________________________

5

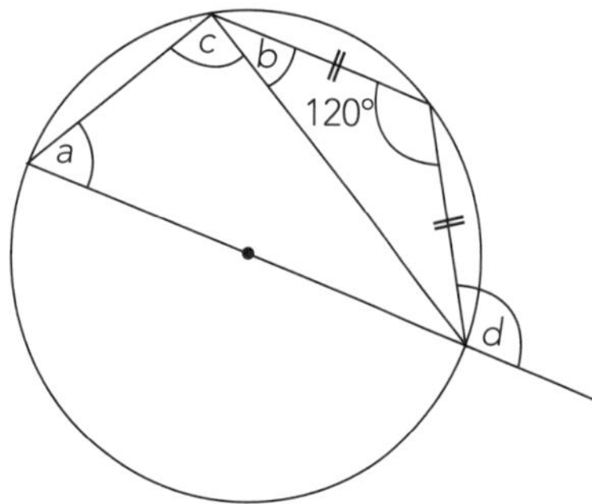

$a =$ ____ ________________________________

$b =$ ____ ________________________________

$c =$ ____ ________________________________

$d =$ ____ ________________________________

ISBN: 9780170370394

6

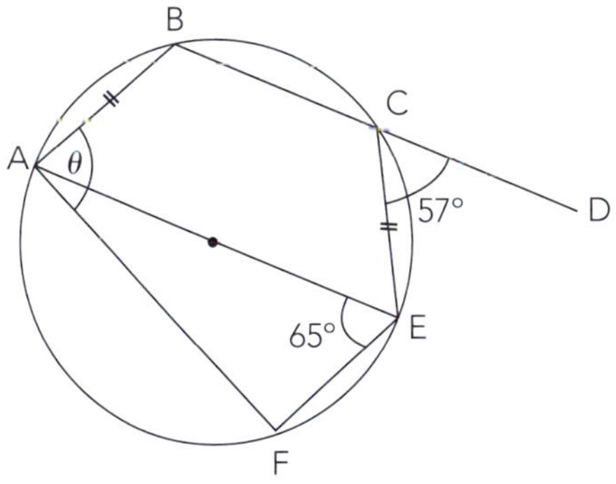

A, B, C, D, E, F, θ, 57°, 65°

7

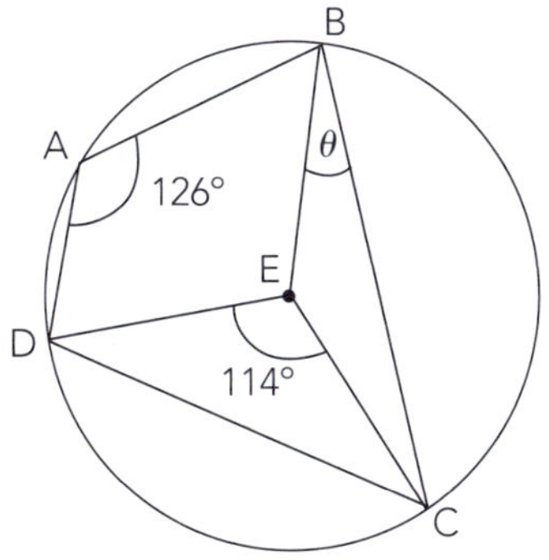

A, B, C, D, E, θ, 126°, 114°

8

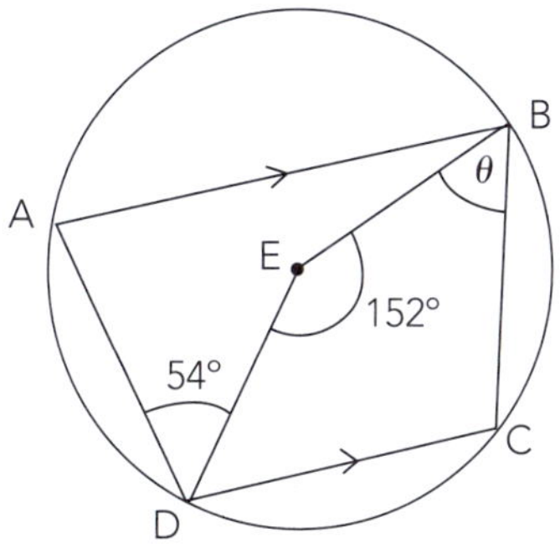

A, B, C, D, E, θ, 152°, 54°

9

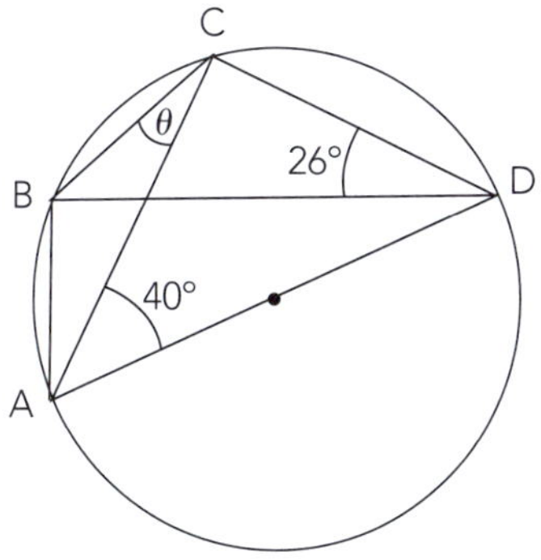

A, B, C, D, θ, 26°, 40°

10

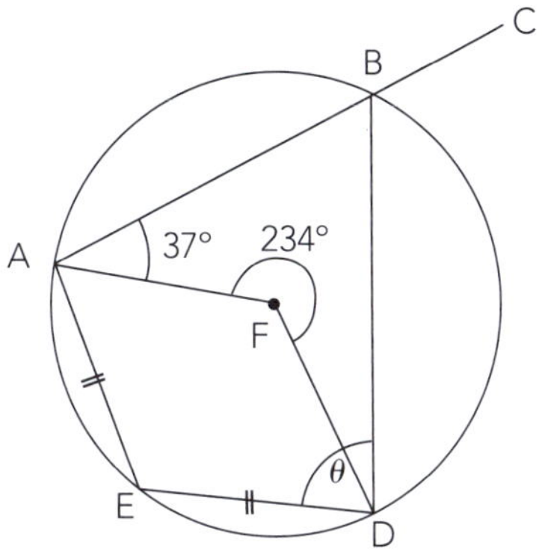

A, B, C, D, E, F, θ, 37°, 234°

Concyclic points

- Concyclic points are points through which a circle can be drawn.
- A circle can always be drawn through **any** three points.

Examples:

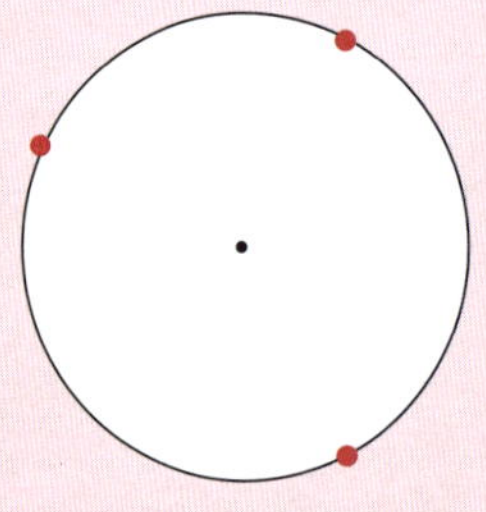

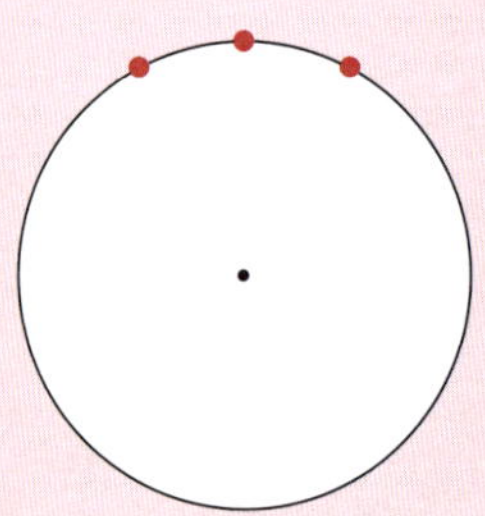

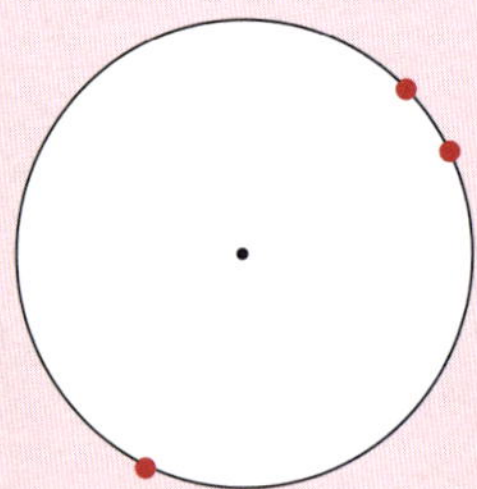

- A circle **cannot** always be drawn through any **four** points.

A circle can be drawn through four points if:

1 Two equal angles (at C and D) extend from the same side of two points (A and B).

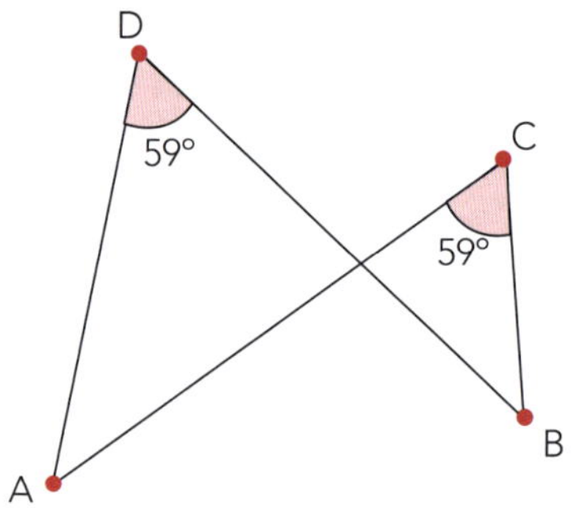

This is because angles on the same arc are equal.

∴ **Points A, B, C and D are concyclic.**

Example: Show that points A, B, C and D are concyclic.

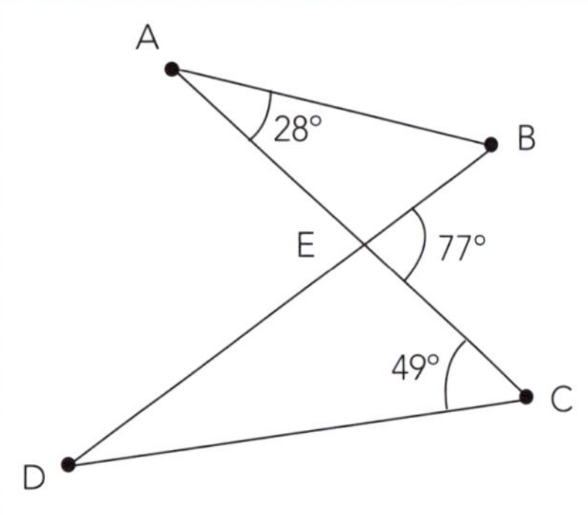

∠DEC = 103° (∠s on a line = 180°)

∠EDC = 28° (∠s in Δ = 180°)

∴ ∠BAE = ∠EDC

∴ Points A, B, C and D are concyclic (∠s on the same arc =)

 ISBN: 9780170370394

2 The opposite angles of a quadrilateral add to 180°.

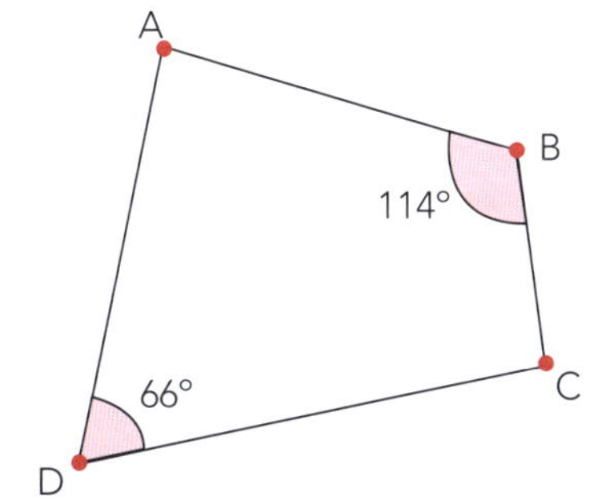

This is because opposite angles of a cyclic quadrilateral add to 180°.

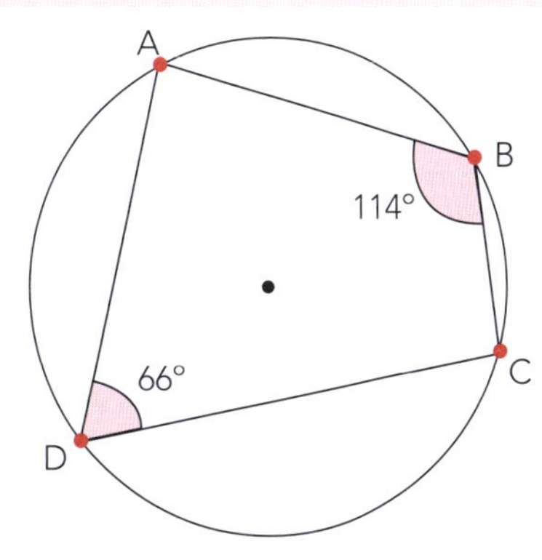

∴ **Points A, B, C and D are concyclic.**

Example one: Is the quadrilateral below cyclic? Justify your answer.

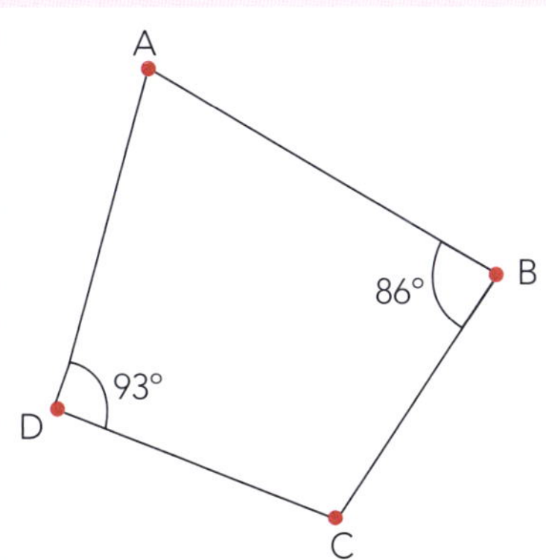

∠ADC + ∠ABC = 179°

∴ Quadrilateral ABCD is **not** cyclic because ∠ADC and ∠ABC do not add to 180°. (opp ∠s of cyclic quad add to 180°)

Example two: Four of the five points below are concyclic. Which are they? Justify your decision.

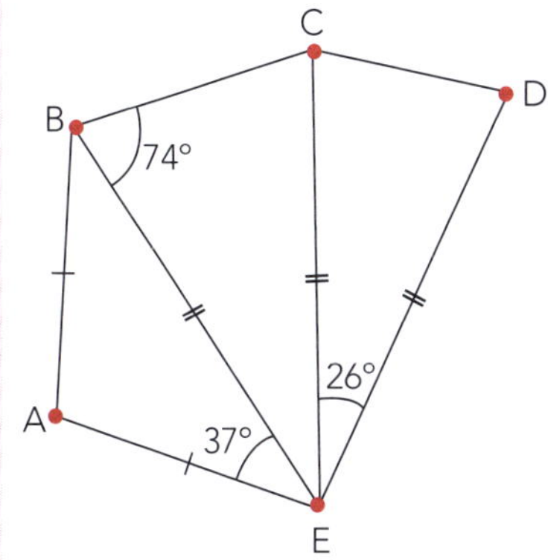

∠BCE = 74° (isos Δ, base ∠s =)
∠EAB = 106° (isos Δ, base ∠s =)
∴ Points A, B, C and E are concyclic because ∠BCE and ∠EAB add to 180°. (opp ∠s of cyclic quad add to 180°)

ISBN: 9780170370394

For each diagram, state whether points A, B, C and D are concyclic, and justify your answer.

1

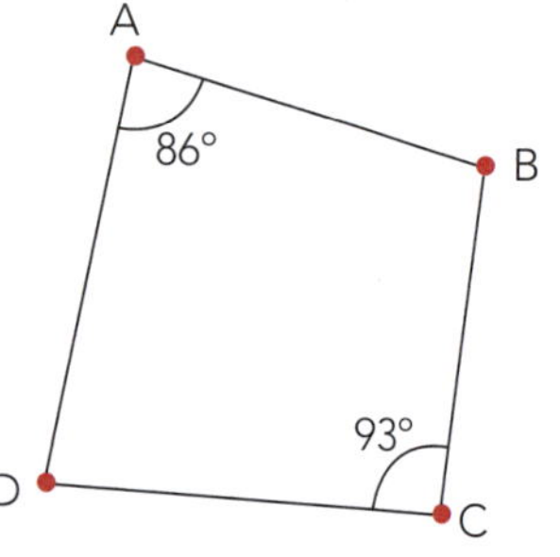

2

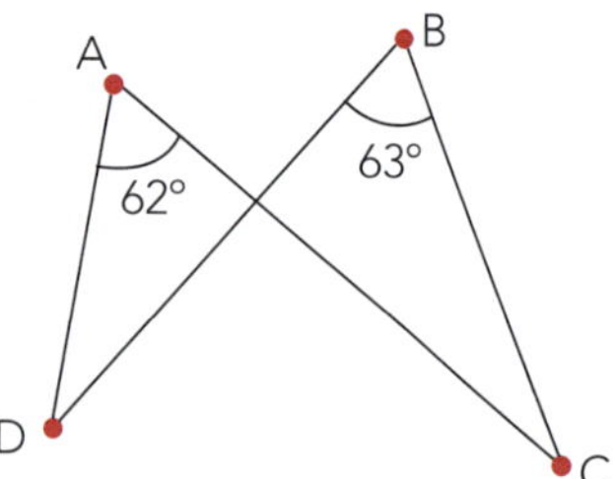

3

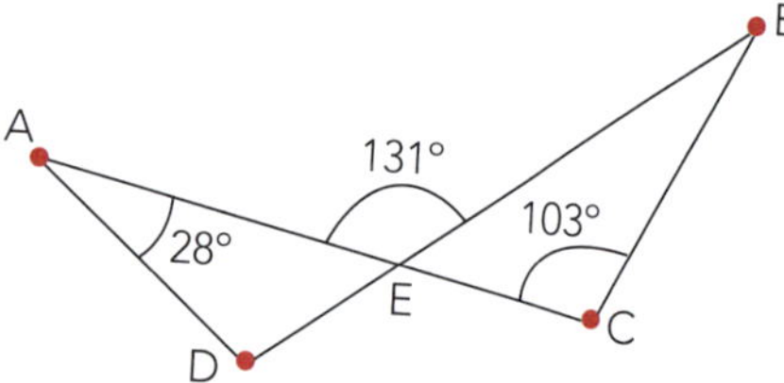

4

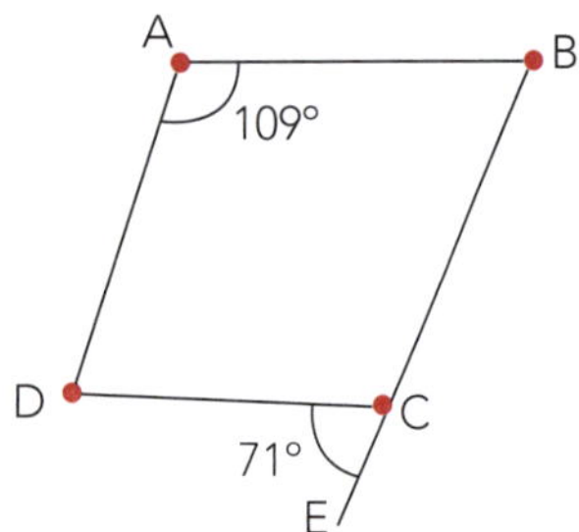

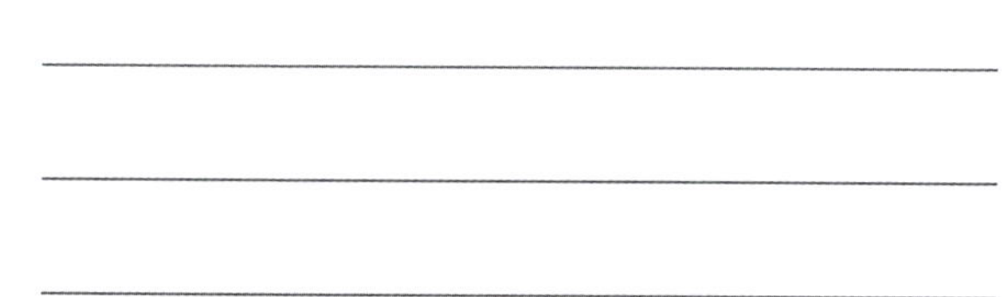

5

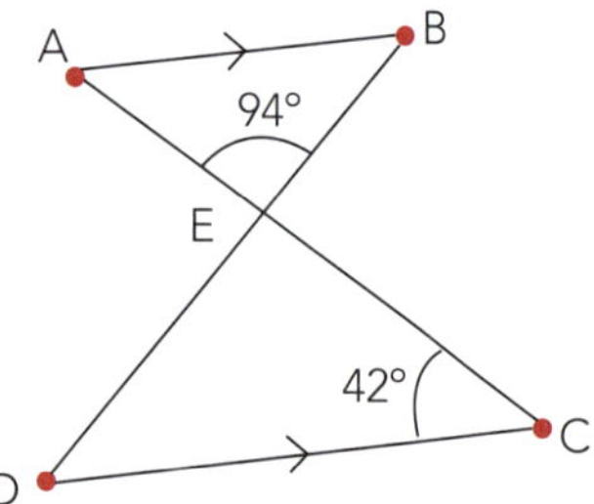

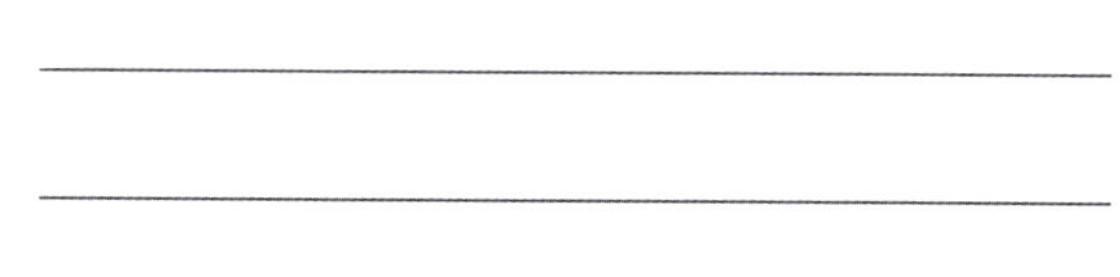

6

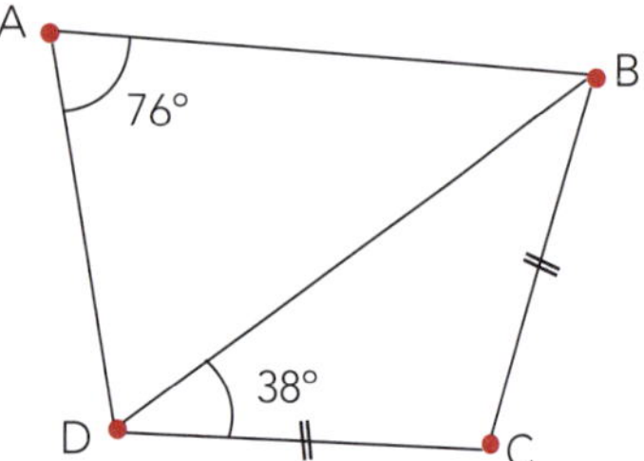

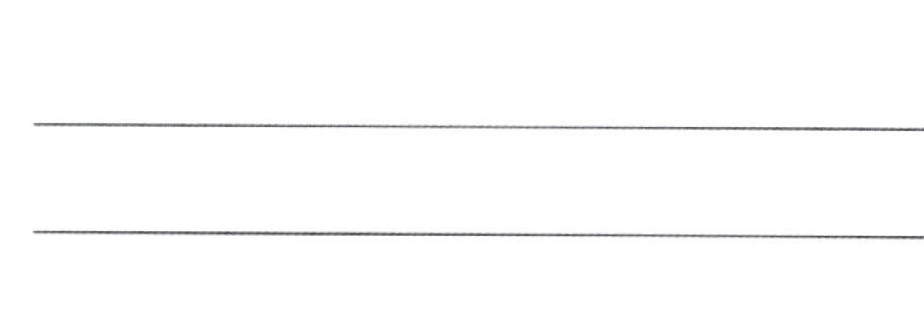

 ISBN: 9780170370394

Four of the points in the following diagrams are concyclic. Identify them and justify your answer.

7

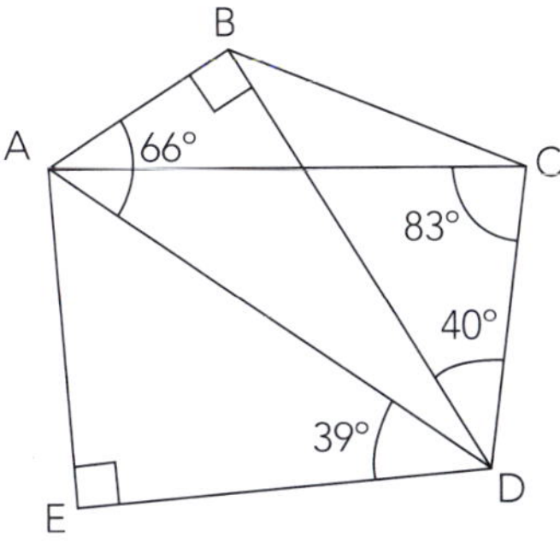

8

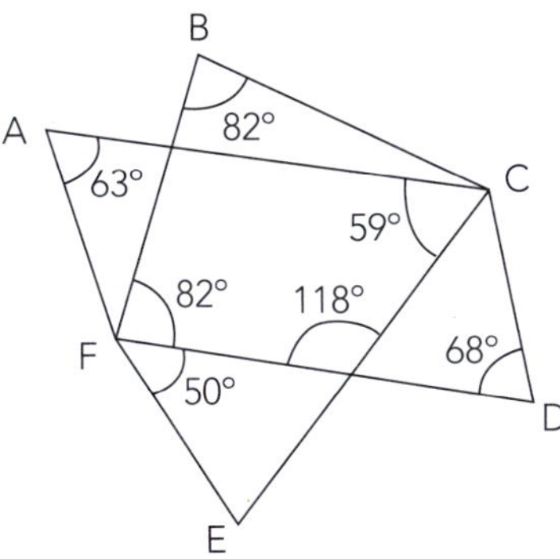

9

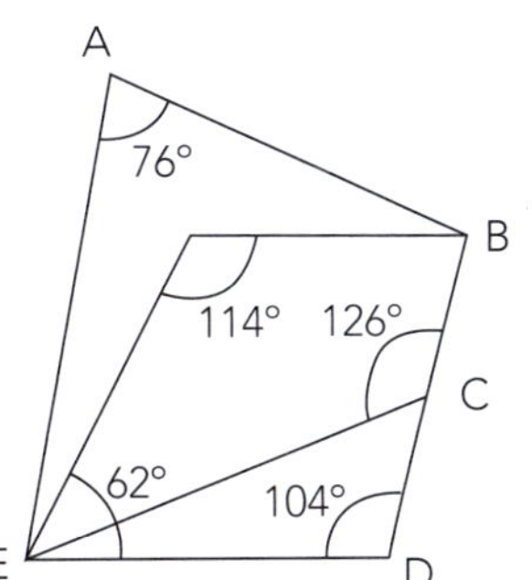

10

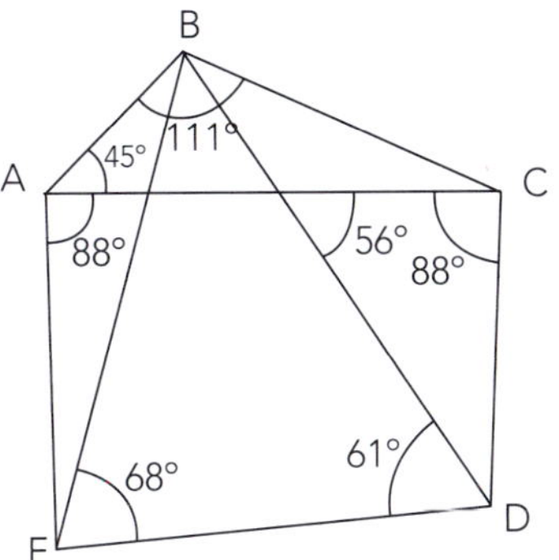

11

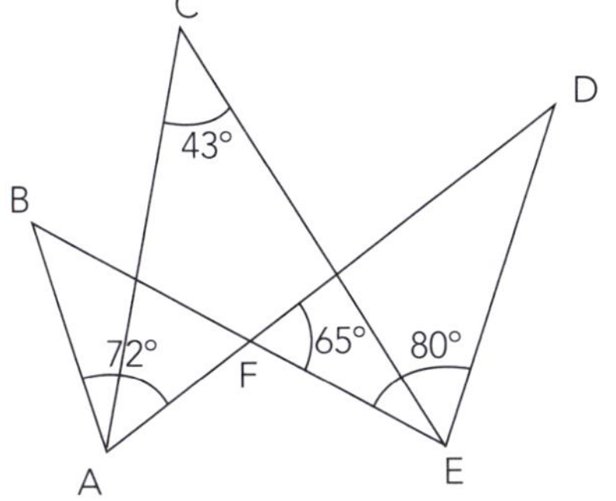

Use of algebra in geometry

- So far you have been asked to find the actual size of an angle.
- You may be asked to find the size of an angle in terms of another angle that is expressed as a variable.

Example one: Find an expression for angle a in terms of x.

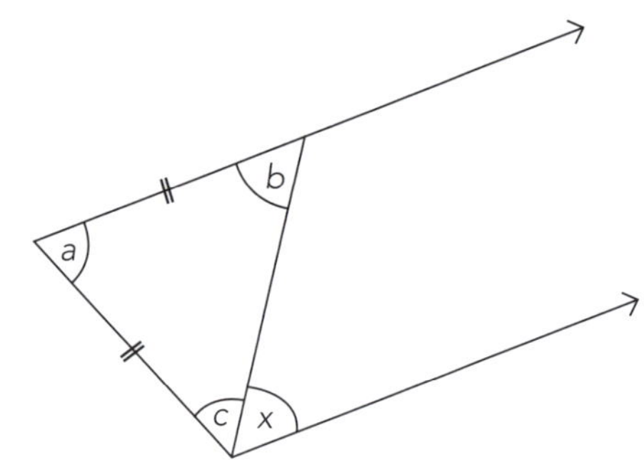

$b = x$ (alt ∠s =, // lines)

$c = x$ (isos Δ, base ∠s =)

∴ $\mathbf{a = 180° - 2x}$ (∠s in Δ = 180°)

Example two: Find the size of ∠ADC in terms of x.

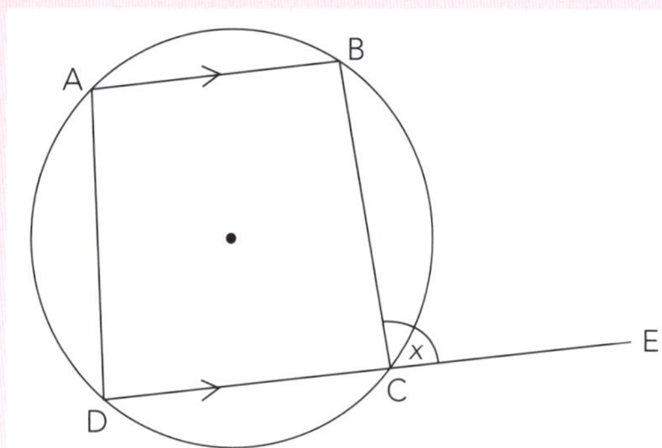

∠BCD = 180° – x
∠ABC = x (alt ∠s =, // lines)
∠BAD = x (ext ∠ of cyclic quad = the int opp ∠)
∴ ∠ADC = 360° – (180° – x) – x – x
∠ADC = 180° – x (∠s of quad = 360° or co-int ∠s add to 180°, // lines)

Example three: Find an expression for the size of angle c in terms of x and y.

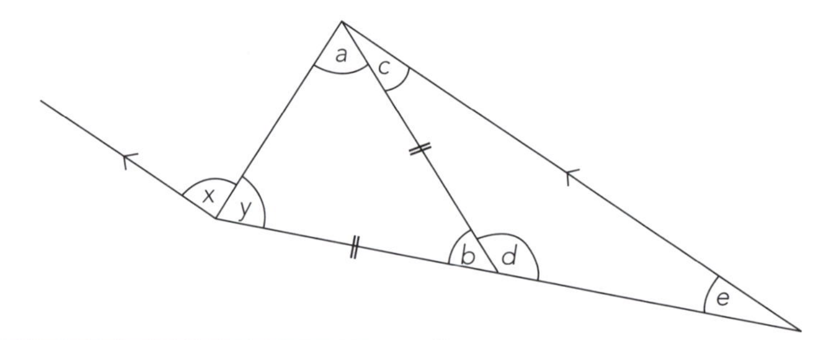

$a = y$ (isos Δ, base ∠s =)
$b = 180° - 2y$ (∠s in Δ = 180°)
$d = 180° - (180° - 2y)$ (∠s on a line = 180°)
$= 2y$
$e = 180° - x - y$ (co-int ∠s add to 180°, // lines)
∴ $c = 180° - 2y - (180° - x - y)$ (∠s in Δ = 180°)
$\mathbf{c = x - y}$

ISBN: 9780170370394

Solve the following problems. In all cases, justify your reasoning.

1 Find an expression for the size of angle b in terms of x.

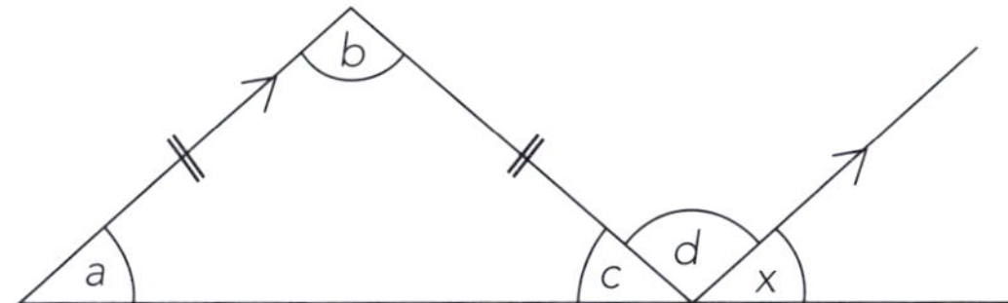

2 Find an expression for the size of angle b in terms of x.

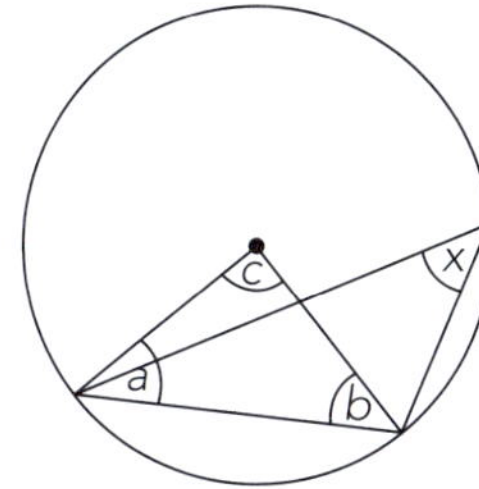

3 Find an expression for the size of $\angle AED$ in terms of x.

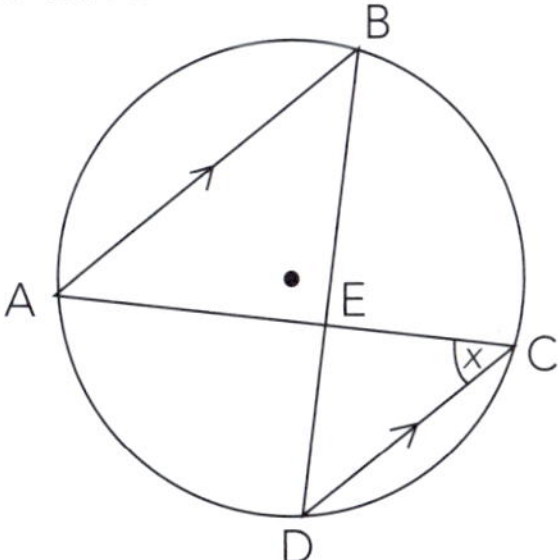

4 Find an expression for the size of $\angle BAD$ in terms of x.

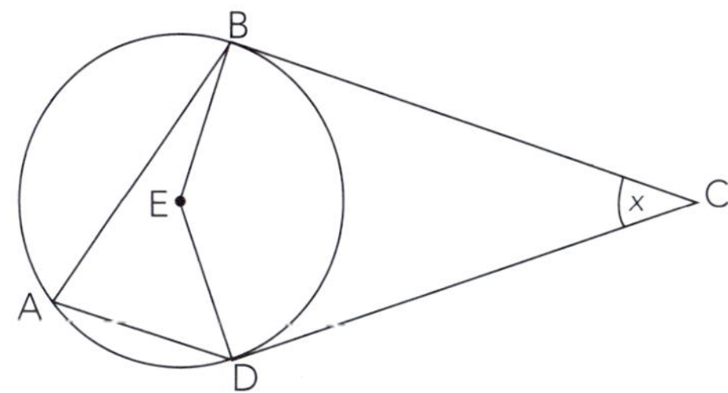

5 Find an expression for the size of angle e in terms of x.

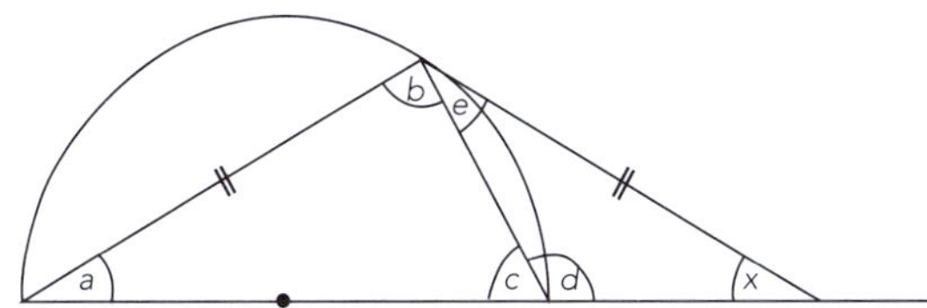

6 Find an expression for the size of angle b in terms of x and y.

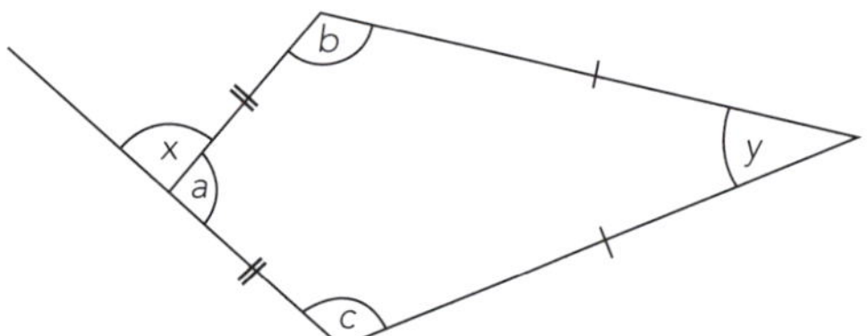

7 Find an expression for the size of angle a in terms of y.

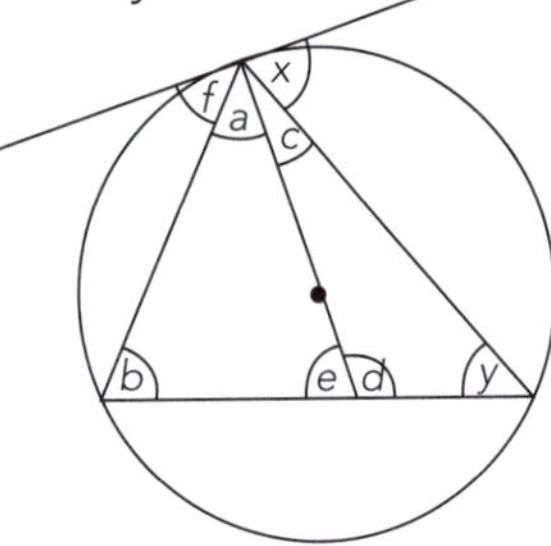

8 Find an expression for the size of angle a in terms of x and y.

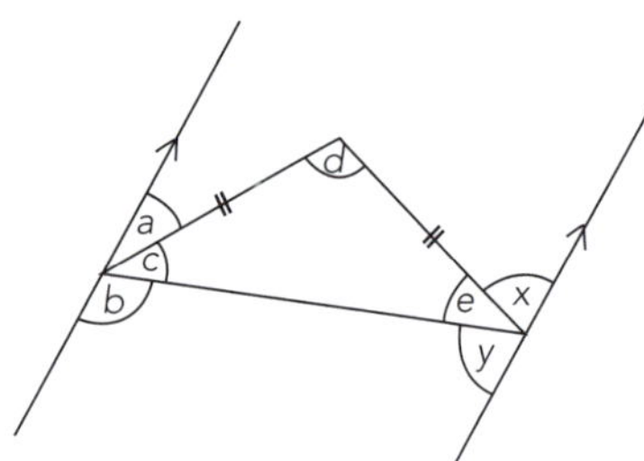

9 Find an expression for the size of ∠BCD in terms of x and y.

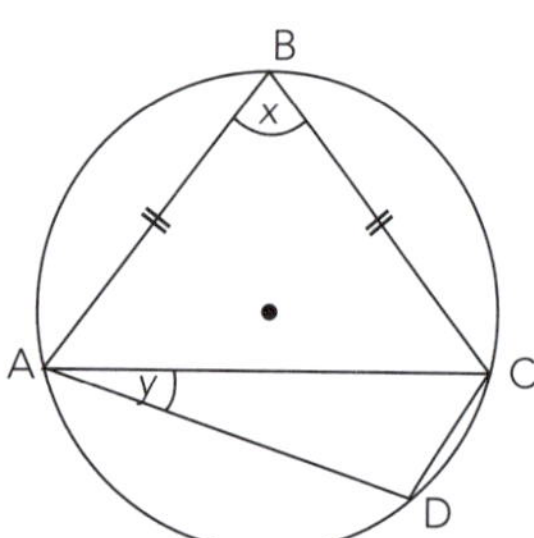

10 Find an expression for the size of angle a in terms of x and y.

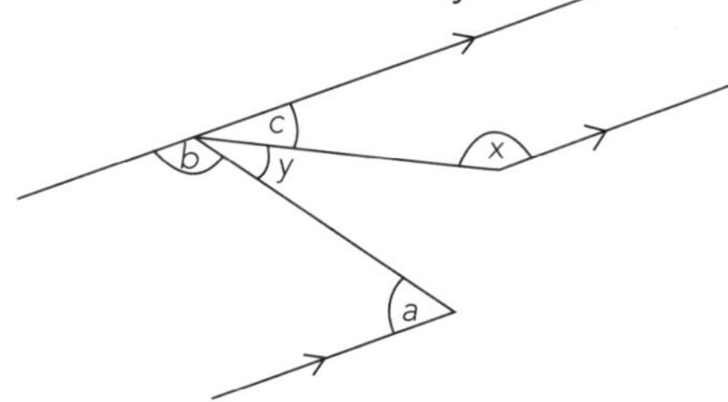

ISBN: 9780170370394

Similar triangles

- Similar triangles are the same shape.
- They have the same-sized angles as each other.

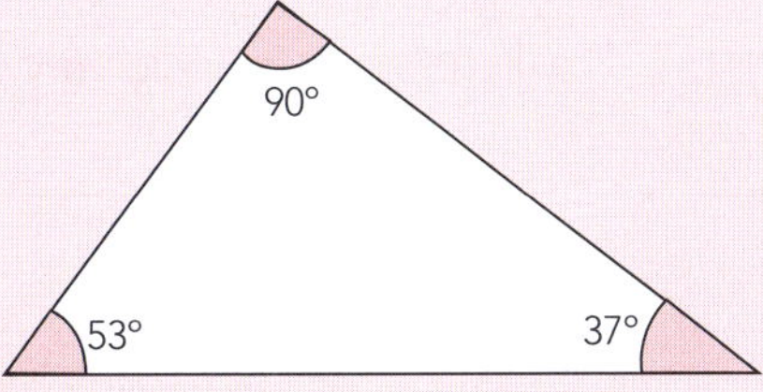

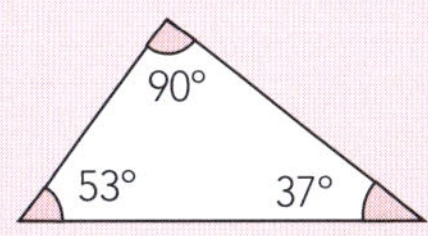

- However, they are not always drawn the same way up.

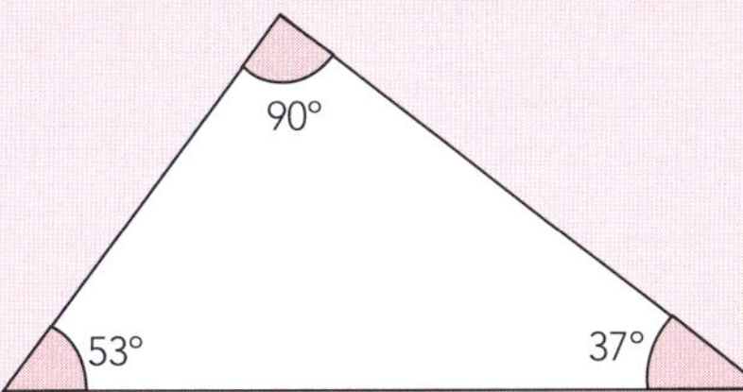

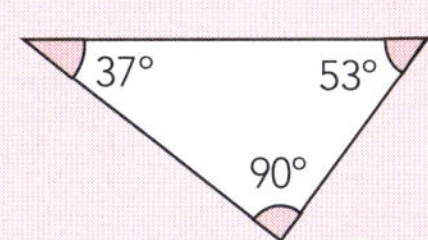

- The angles can also be in reverse order.

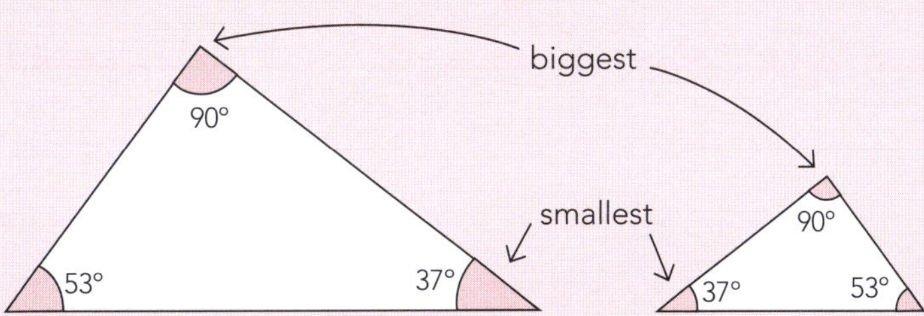

- The sides are also in proportion.

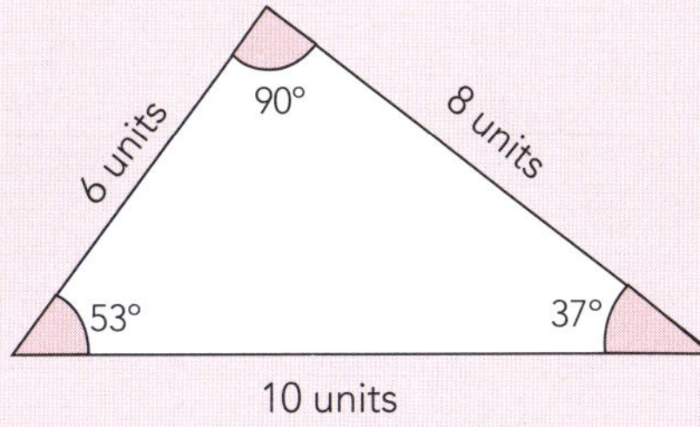

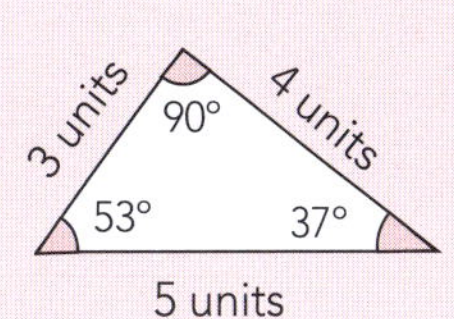

10 units

5 units

In this case: $\dfrac{\text{big triangle}}{\text{small triangle}} = \dfrac{10}{5} = \dfrac{8}{4} = \dfrac{6}{3} = 2$

- It is customary to label the vertices in corresponding order.

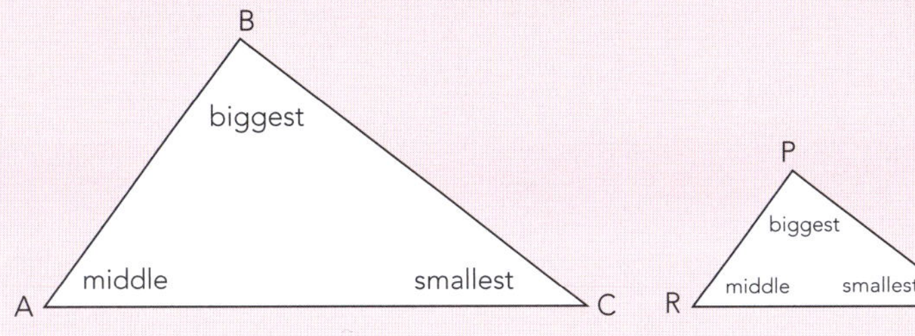

In this case: Δ**BAC** is similar to Δ**PRQ**.

Both are in **biggest, middle, smallest** order

Examples using angles:
Write a statement by each pair of triangles. Justify your answers.

1

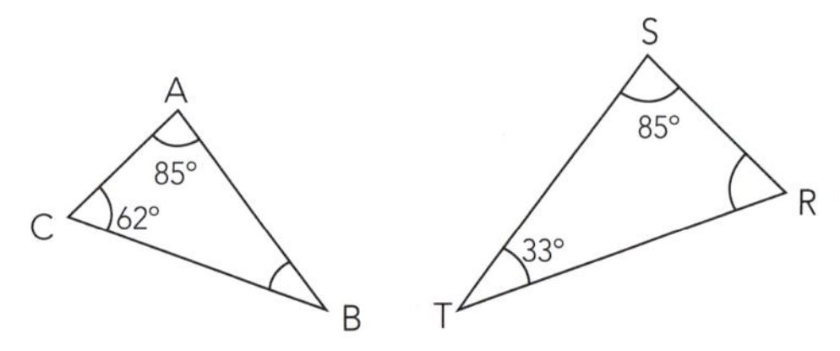

∠ABC = 33° (∠s in Δ = 180°)

∴ ΔABC is similar to ΔSTR because the angles are equal.

biggest, smallest, middle

2

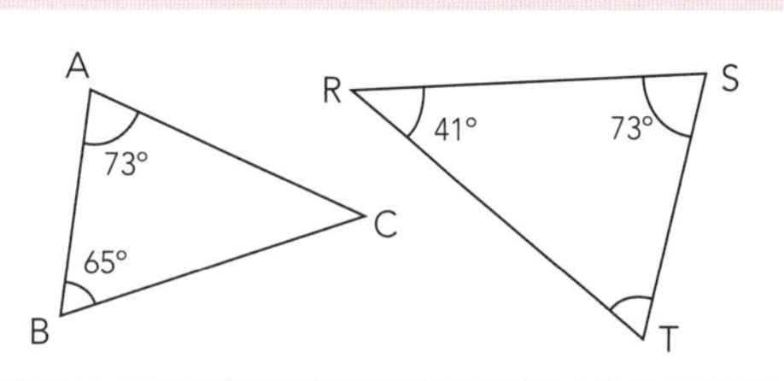

∠ACB = 42° (∠s in Δ = 180°)

∴ ΔABC is not similar to ΔSTR because the angles are not equal.

State whether the following pairs of triangles are similar or not, and justify your answer.

1

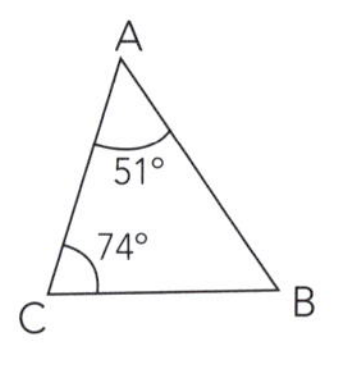

D, 51°, 74°, F, E

____
____
____

2

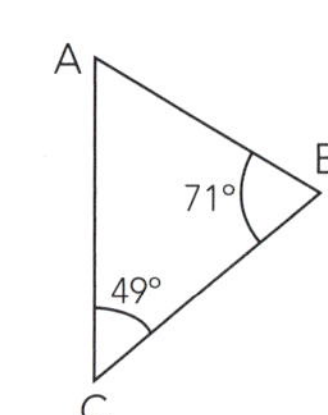

D, 49°, 60°, E, F

____
____
____

3

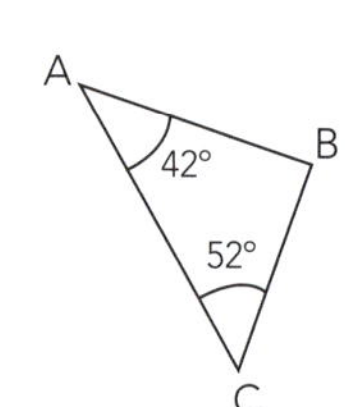

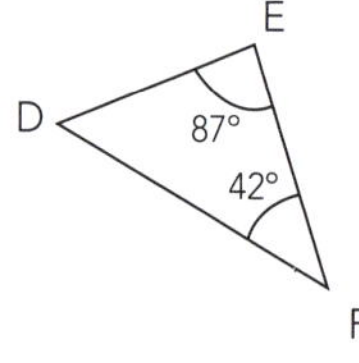

____
____
____

4

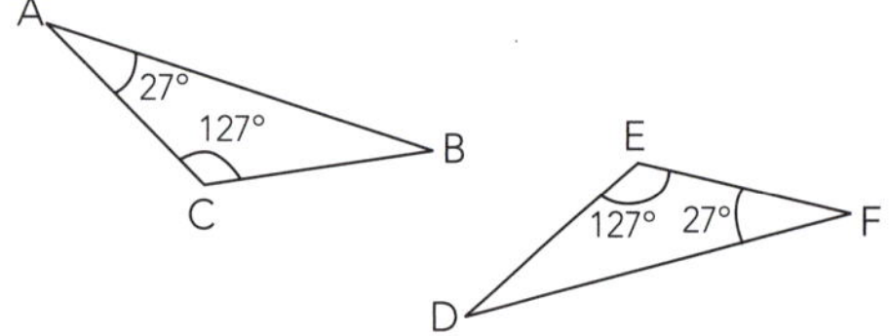

____
____
____

5 Triangles ABC and DEB

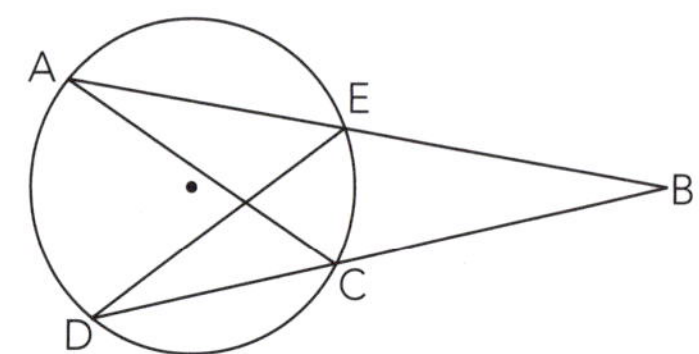

____
____
____

 ISBN: 9780170370394

Examples using sides:

Write a statement by each pair of triangles. Justify your answers.

1

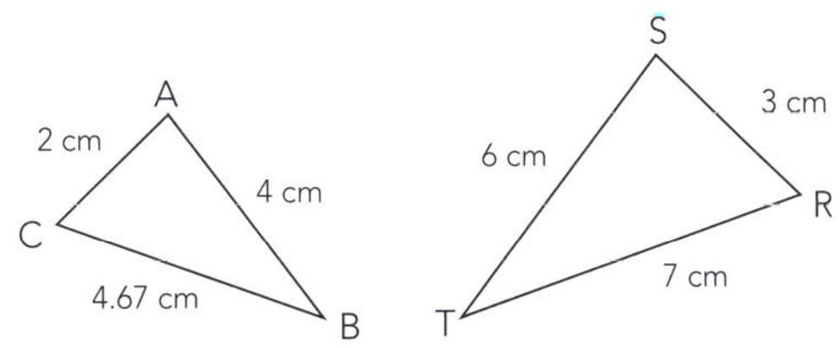

$$\frac{2}{3} = \frac{4}{6} = \frac{4.67}{7} = 0.67 \text{ (2 dp)}$$

$$\textbf{OR } \frac{3}{2} = \frac{6}{4} = \frac{7}{4.67} = 1.5 \text{ (2 dp)}$$

$\therefore$ Δ**ABC** is similar to Δ**STR** because the sides are in proportion.

biggest, smallest, middle

2

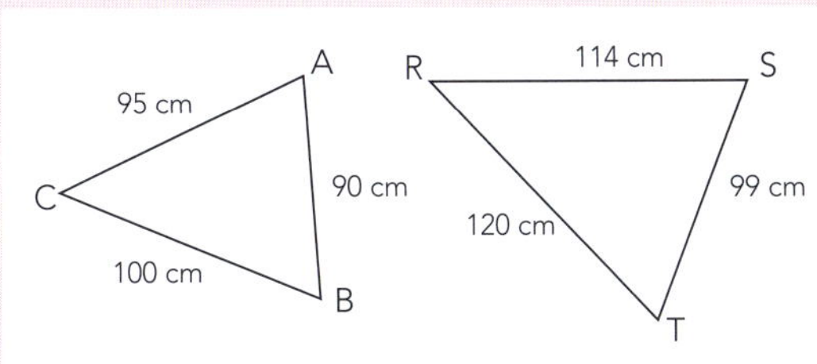

$$\frac{120}{100} = \frac{114}{95} = 1.2 \qquad \frac{99}{90} = 1.1$$

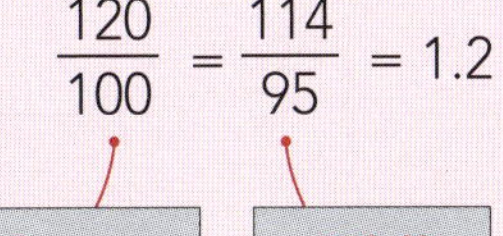

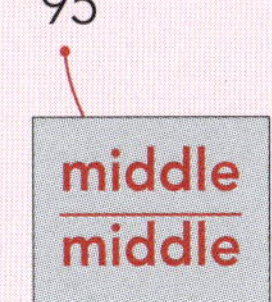

biggest / biggest — **middle / middle** — **smallest / smallest**

$\therefore$ Δ**ABC** is not similar to Δ**STR** because the sides are not in proportion.

State whether the following pairs of triangles are similar or not, and justify your answer.

6

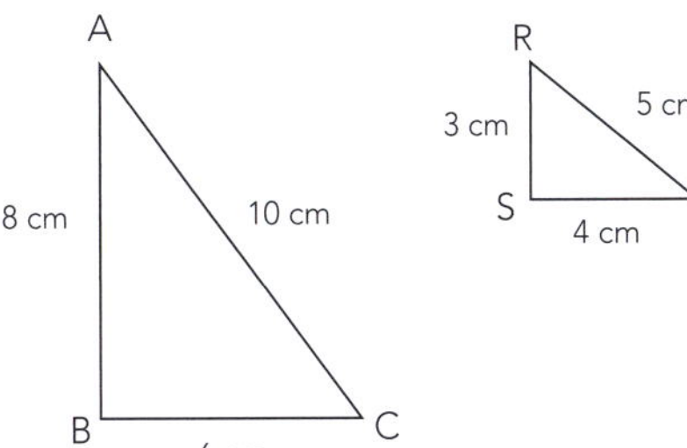

____

____

____

7

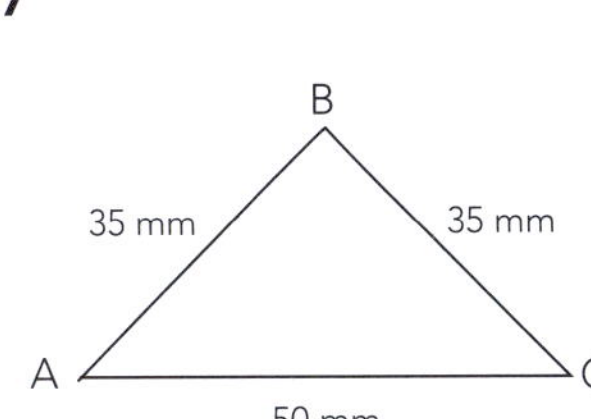

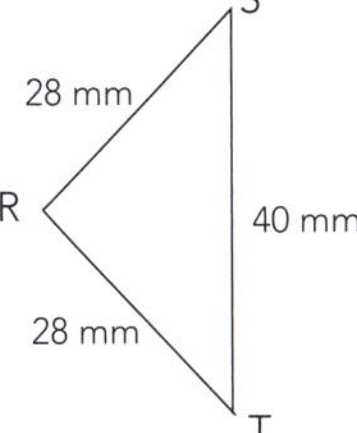

____

____

____

8

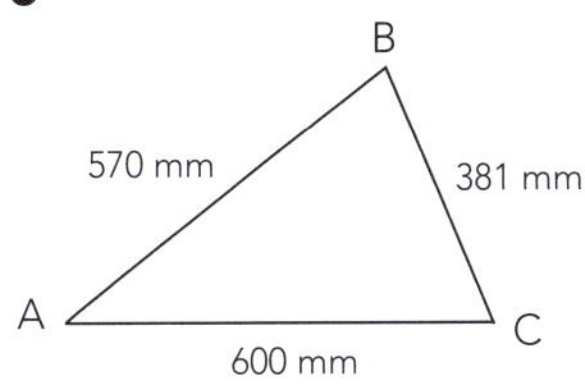

S, 254 mm, 395 mm, R, T, 400 mm

____

____

____

ISBN: 9780170370394

Calculating unknown sides

Because the sides of similar triangles are in proportion, if we know one pair of corresponding sides (e.g. both the longest sides) and we are given the length of another side (e.g. a shortest side), we can calculate the length of its corresponding side.

Example: Calculate the side x.

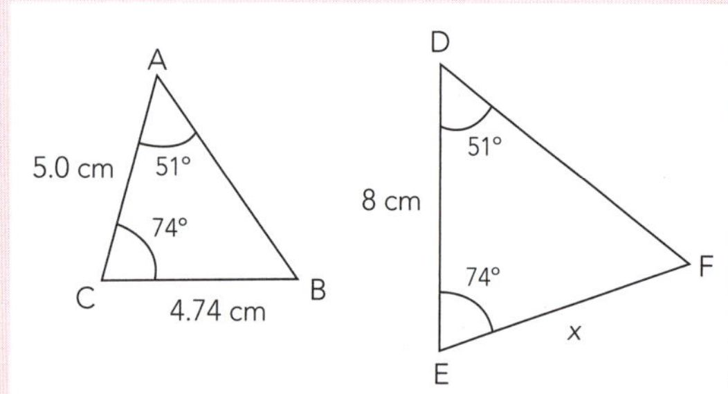

ΔABC is similar to ΔDFE ($\angle$s in Δ = 180°)

$$\therefore \frac{AC}{DE} = \frac{CB}{EF}$$

$$\frac{5.0}{8.0} = \frac{4.74}{x}$$

$$5.0x = 8.0 \times 4.74$$

$$x = 7.584 \text{ cm}$$

Explain why triangles in the following figures are similar and calculate the value of x.

9

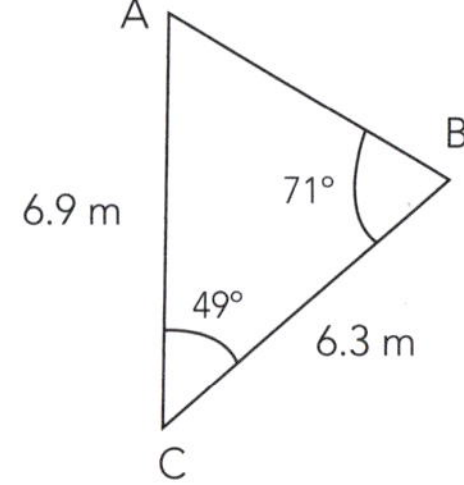

D, 7.2 m, 49°, x, E, 60°, F

10

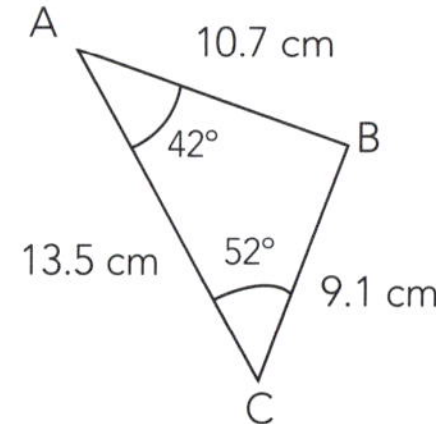

E, D, 52°, x, 42°, 6.8 cm, F

11

B, 3.5, C, 71°, x, 61°, A

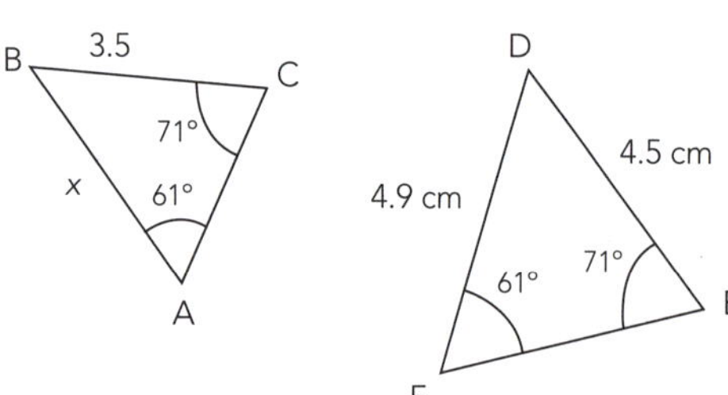

12

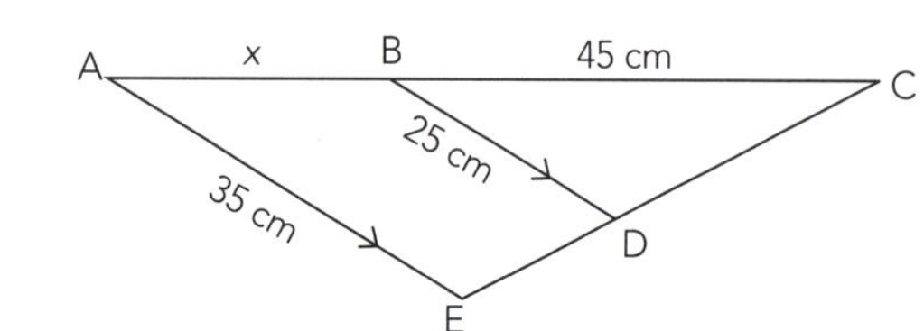

 ISBN: 9780170370394

13

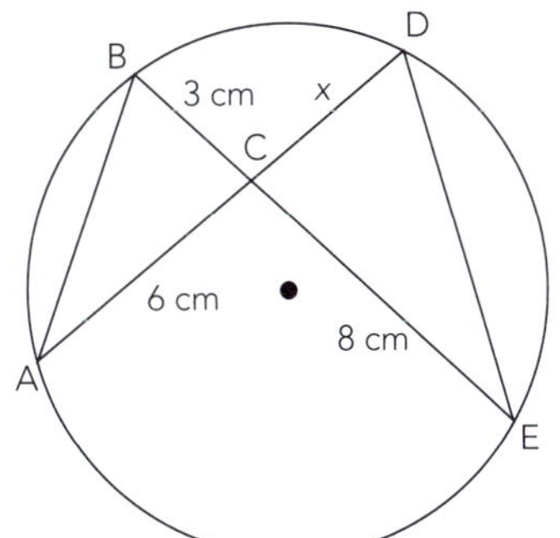

14

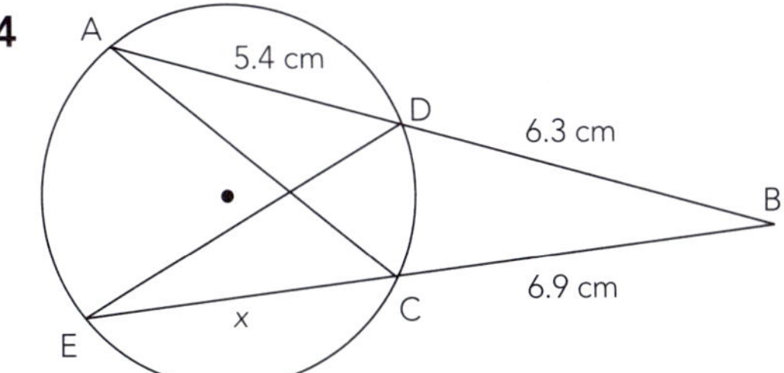

15

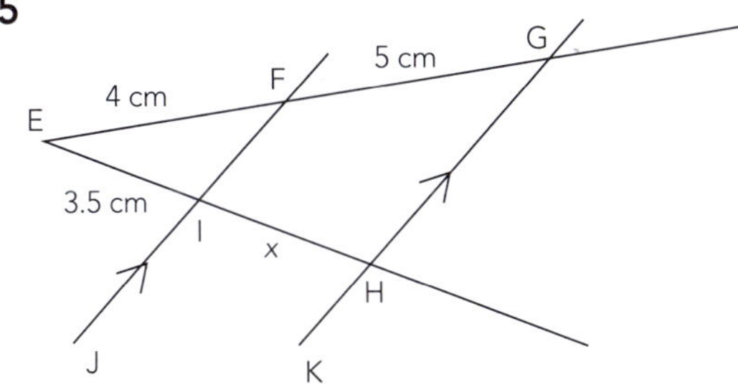

16

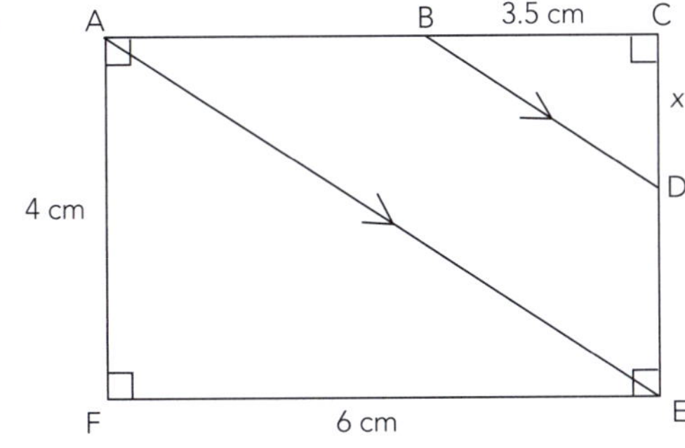

17

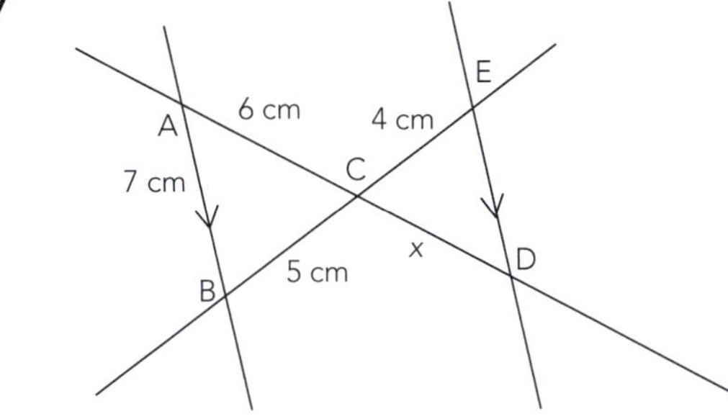

ISBN: 9780170370394

Proof

- A proof is a mathematical argument.
- You need to give geometric reasons for each step.
- Proofs must be general, so they use letters rather than numbers.
- Sometimes devising a proof requires the construction of extra lines.
- It can also be useful to name angles x, y, etc.
- Frequently, a previous question will hint at how a proof can be done.

Example one: Prove that the diagonals of a rhombus intersect at right angles.

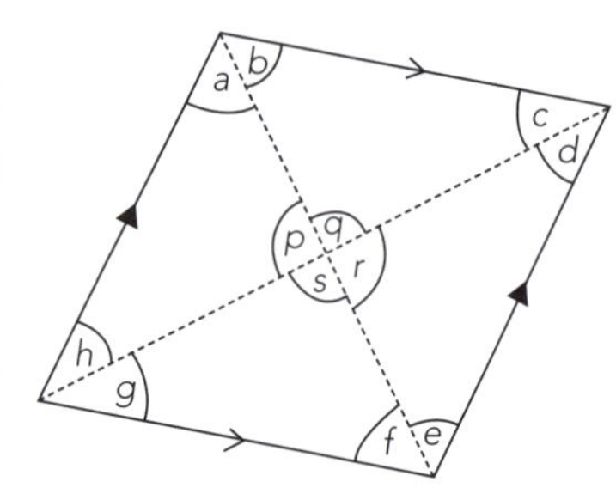

$\angle b = \angle e$ (isos θ, base $\angle$s =)
$\angle b = \angle f$ (alt $\angle$s =, // lines)
$\therefore\ \angle e = \angle f$

$\angle h = \angle c$ (isos θ, base $\angle$s =)
$\angle h = \angle d$ (alt $\angle$s =, // lines)
$\therefore\ \angle c = \angle d$

You do not need to repeat your argument — use the word '**similarly**'

$(c + d) + (e + f) = 180°$ (co-int $\angle$s add to 180°, // lines)
$\therefore\ d + e = 90°$
$\therefore\ r = 90°$ ($\angle$s in Δ = 180°)
Similarly, *s*, *p* and *q* equal 90°.

Example two: Prove that **angle A + angle C = 180°**

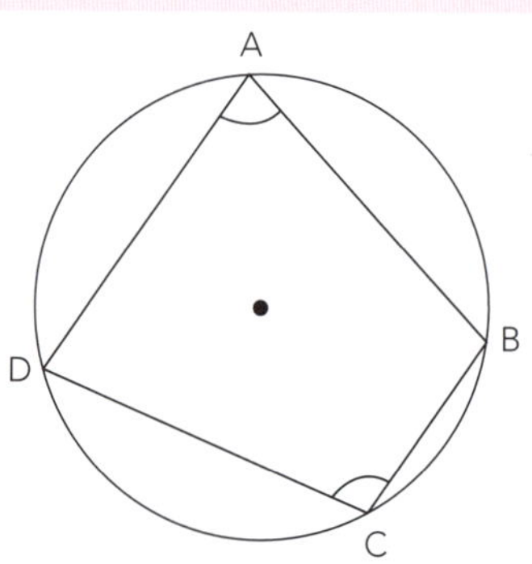

Step 1: Construct lines linking D and B to the centre of the circle, and label the angles at the centre *x* and *y*.

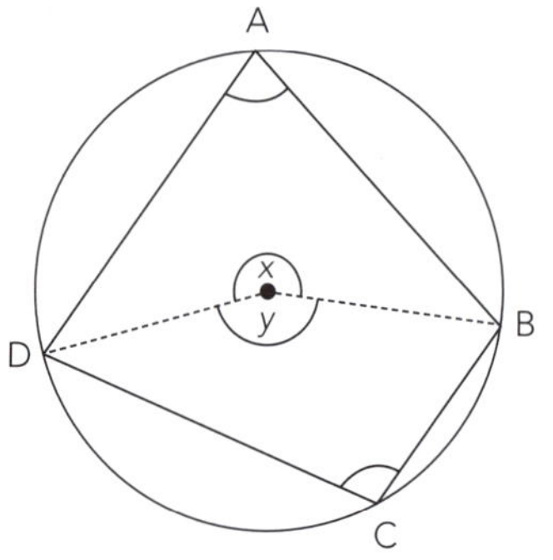

Step 2: $\angle x = 2 \times \angle C$ ($\angle$ at cent is 2 x $\angle$ at circ)
$\angle y = 2 \times \angle A$ ($\angle$ at cent is 2 x $\angle$ at circ)

But $\angle x + \angle y = 360°$ ($\angle$s at a point = 360°)
$\therefore\ 2 \times \angle C + 2 \times \angle A = 360°$
$\therefore\ \angle C + \angle A = 180°$

ISBN: 9780170370394

Fill in the gaps to complete the following proofs.

1 Prove that $\angle$**BCD =** $\angle$**ABC +** $\angle$**EDC**

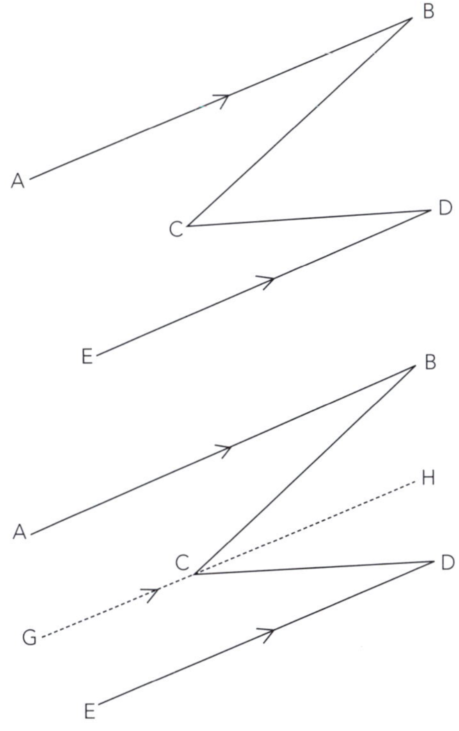

Step 1: Construct a line parallel to AB and ED that passes through C. Call this GH.

Step 2: Use parallel line rules.

$\angle$ABC = $\angle$BCH (____________________)

$\angle$EDC = $\angle$ _____ (____________________)

$\angle$BCD = $\angle$BCH + $\angle$ _____

∴ $\angle$**BCD =** $\angle$**ABC +** $\angle$**EDC**

2 Prove that **CA x CB = CE x CD**

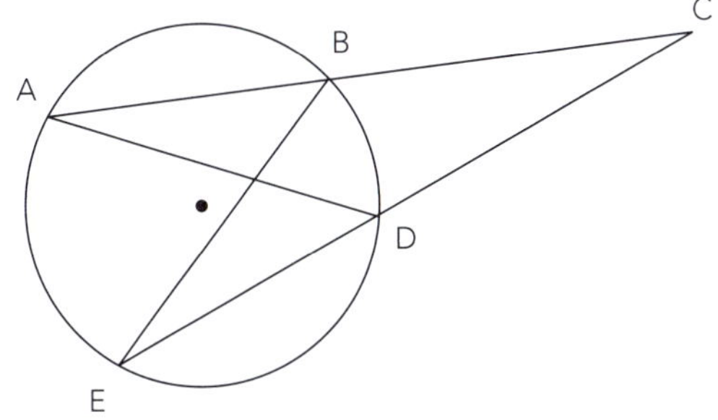

Step 1: Show that triangles CAD and CEB are similar.

$\angle$C is common to both triangles.

$\angle$CAD = $\angle$ _____ ($\angle$s on the same arc =)

∴ $\angle$CDA = $\angle$ _____ ($\angle$s in Δ = 180°)

∴ ΔCAD is ____________________ to ΔCEB

Step 2: Because the triangles are

____________________, their sides are in

____________________.

ΔCAD is ________________ to ΔCEB so:

$$\frac{CA}{CE} = \frac{CD}{\_\_\_\_}$$

∴ **CA x CB = CE x CD**

ISBN: 9780170370394

Complete the following proofs.

3 Prove that **the external angle of a triangle = the sum of the internal opposite angles (∠c = ∠a + ∠b). (Hint: You will need to add a line.)**

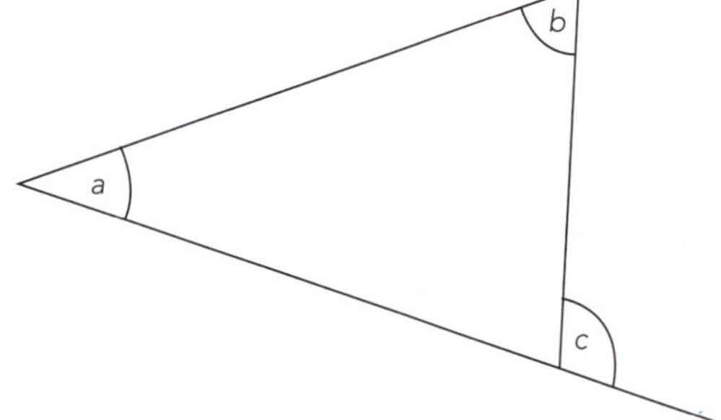

4 Prove that **BA² = BC x BD**.

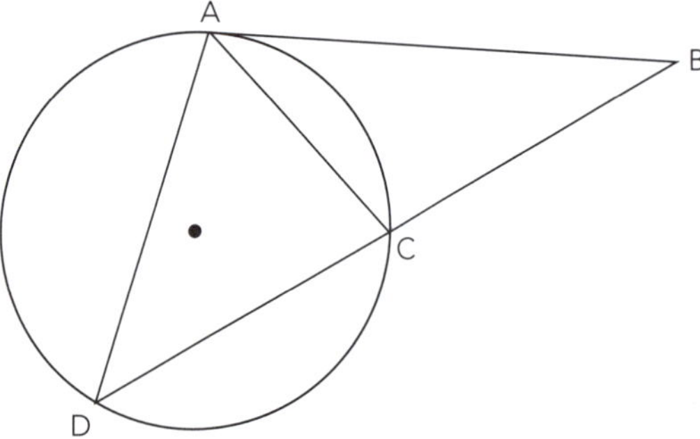

ISBN: 9780170370394

5 Prove that **the angle at the centre = twice the angle at the circumference (∠i = 2(∠b + ∠c))**.

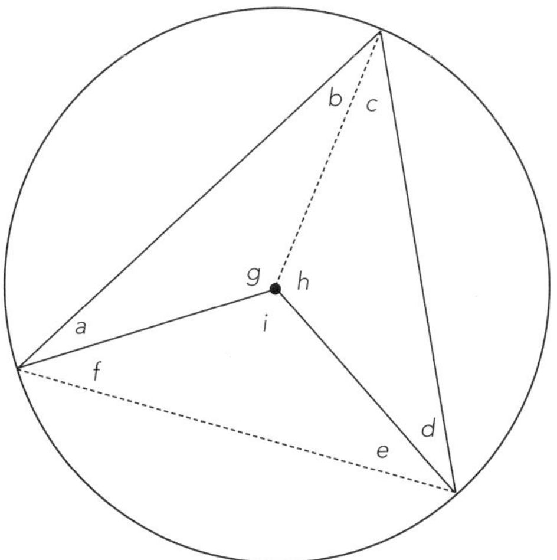

6 Prove that **the angle in a semicircle = 90° (b + c = 90°)**.

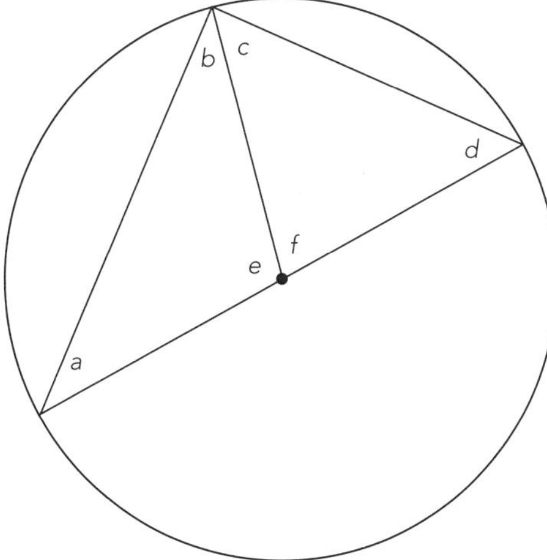

ISBN: 9780170370394

Practice questions

1a **i** Find the size of angle x. Explain your method clearly, and give geometric reasons for each step.

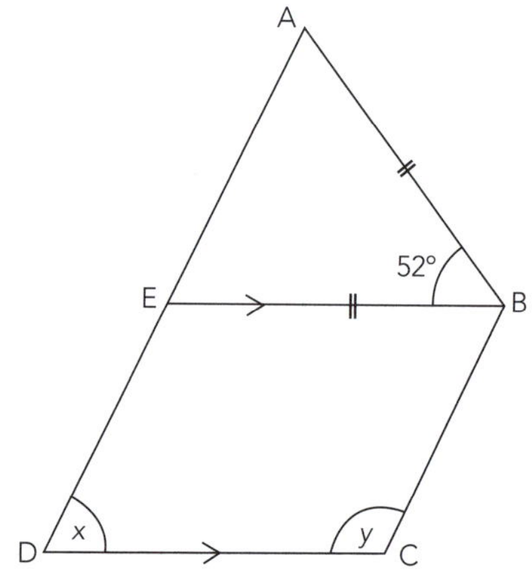

ii William measures angle y and finds it is 116°. He says that this means that figure ABCD must be a trapezium. Is he correct? Explain your method clearly, and give geometric reasons for each step.

iii Find an expression for the size of angle g in terms of angle f. Explain your method clearly, and give geometric reasons for each step.

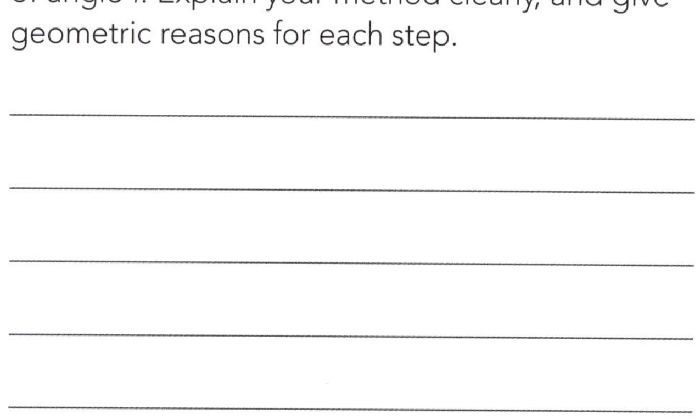

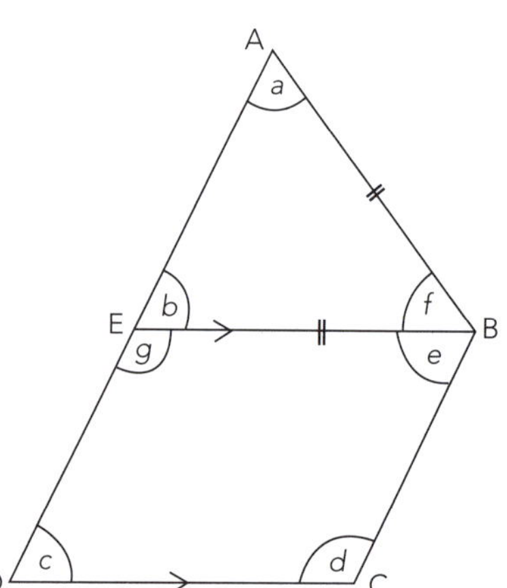

ISBN: 9780170370394

b Figure ABCDE is a regular pentagon. CDF and AEF are straight lines. G is the centre of the pentagon.

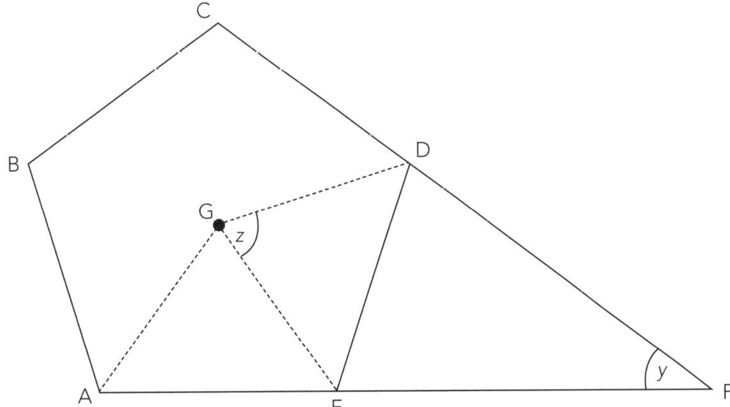

i Find the size of angle *y*. Explain your method clearly, and give geometric reasons for each step.

ii Show that angle $z = 72°$. Explain your method clearly, and give geometric reasons for each step.

iii Marama suggests that figure AGDF is a cyclic quadrilateral. Is she correct? Explain your method clearly, and give geometric reasons for each step.

ISBN: 9780170370394

2a A cross-country course goes north (000°) for 1700 m, east (090°) for 1200, and back to the start.

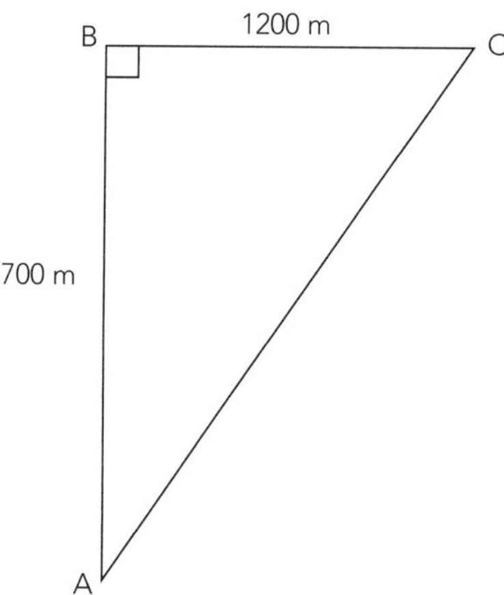

i Calculate the length of the course.

ii Explain why the bearing of A from C is 215.2°.

b The kids' cross-country running course has three stages. The first (AB) is 720 m long on a bearing of 124°. The second (BC) is 540 m long and on a bearing of 34°. The third stage (CA) returns to the start.

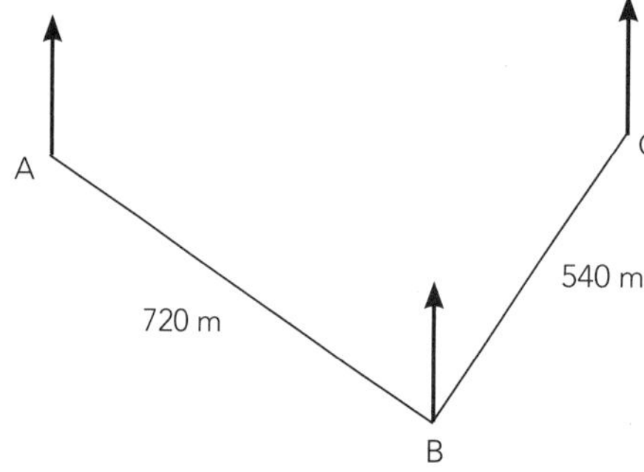

i Calculate the length of the course.

ii Calculate the bearing from C back to the start of the course.

  ISBN: 9780170370394

iii The race organiser wants to construct a longer route for the older runners. He decides to add two stages between C and A. They will both be 600 m long, extend to the north of AC and will be called CD and DA.
Calculate the bearing from C to D.

c The previous year's course had a different design. The first stage (AB) was 620 m long on bearing of 113°. The second stage (BC) was on a bearing of 023°. The third stage (CA) returned to the start, and had a bearing of 241°.
Calculate the length of the course. Use the space to sketch a diagram. Explain your method clearly, and give geometric reasons for each step.

ISBN: 9780170370394

3a i Explain why angle x = angle y.

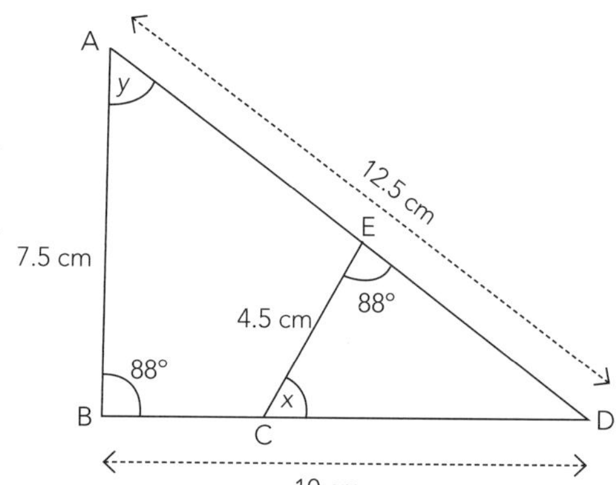

ii Calculate the length of CD.

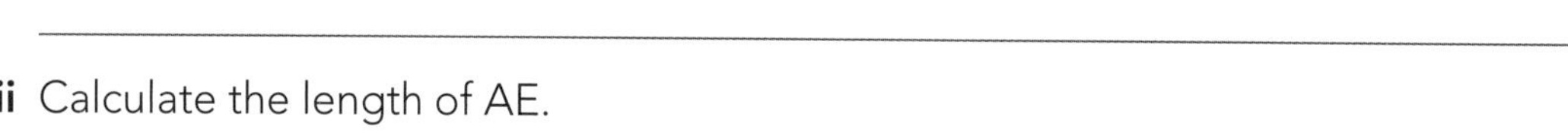

iii Calculate the length of AE.

b i Find the size of angle x. Explain your reasoning.

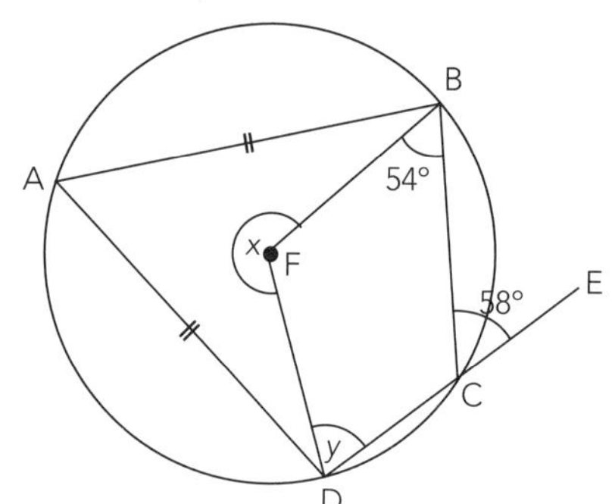

ii Find the size of angle y. Explain your reasoning.

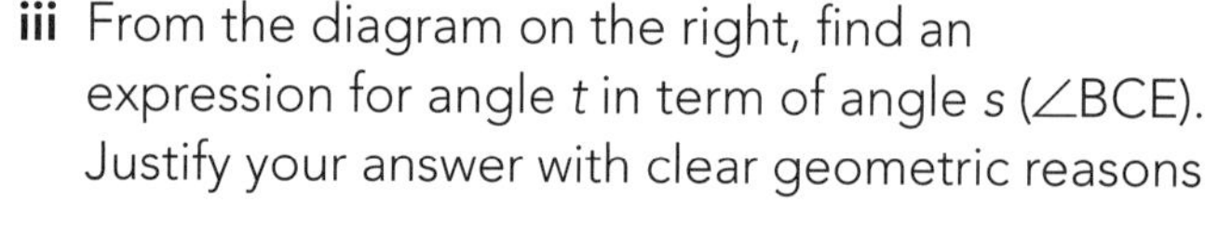

iii From the diagram on the right, find an expression for angle t in term of angle s (∠BCE). Justify your answer with clear geometric reasons.

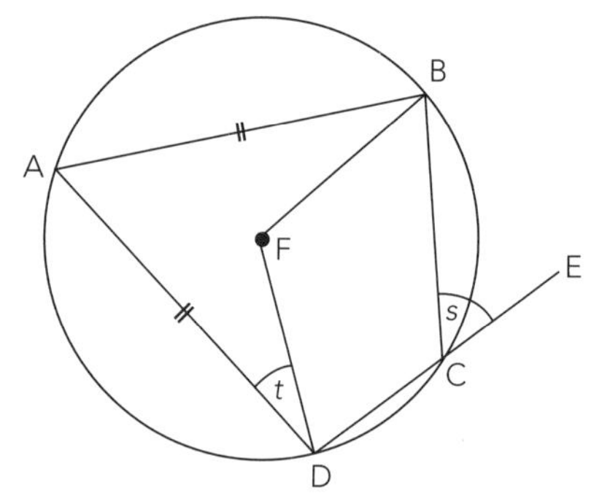

 ISBN: 9780170370394

c AC is a tangent to the circle. DB is a radius. DF bisects triangle DEB.

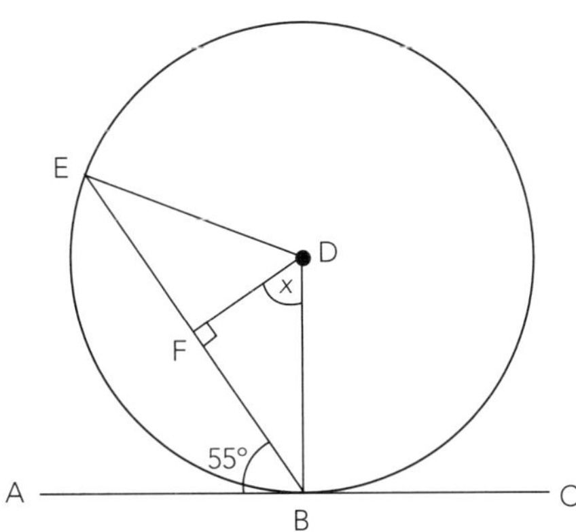

i Aroha thinks that angle $x = 55°$. Is she correct? Give geometric reasons for your answer.

ii Prove that $\angle p = \angle q$. Explain your method clearly, and give geometric reasons for each step.

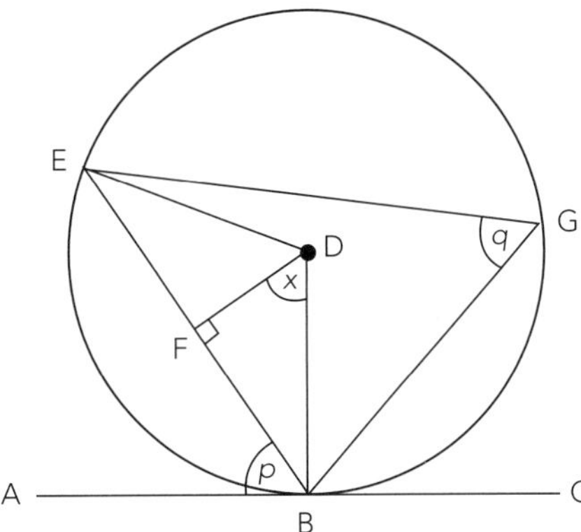

ISBN: 9780170370394

Answers

All non-integral angles are rounded to 1 dp.
All non-integral intermediate answers are rounded to a maximum of 4 sf.
All non-integral final answers are rounded to a maximum of 4 sf.
In many cases involving geometric reasoning, there are alternative correct answers.

The Theorem of Pythagoras (pp. 6–11)

Finding the length of the hypotenuse (pp. 6–7)

1 $x = 19.21$ m
2 $c = 9.434$ m
3 $x = 13$ cm
4 $x = 12.08$ cm
5 $x = 6.940$ m
6 $x = 42.43$ mm
7 $L = 91.53$ mm
8 $P = 104.8$ m

Finding the lengths of short sides (p. 8)

1 $x = 16.94$ mm
2 $x = 48.99$ mm
3 $x = 7.483$ cm
4 $L = 15.49$ km
5 $L = 45.92$ mm
6 $P = 0.5905$ km

Mixing it up (p. 9)

1 50.99 cm
2 122.1 mm
3 12 cm
4 6.788 cm
5 4.272 km
6 68.23 mm
7 61.81 cm
8 128.7 km
9 363.0 mm
10 0.5318 cm

Applications (pp. 10–11)

1 AC = 21.21 cm, AE = 10.61 cm
2 AB = 92.65 mm
3 AB = 21.02 cm
4 BC = 8 cm
5 Perimeter = 48.28 cm
6 $L = 3.94$ m
7 $H = 3.747$ m
8 AB = 9.192 cm

Trigonometry (pp. 12–24)

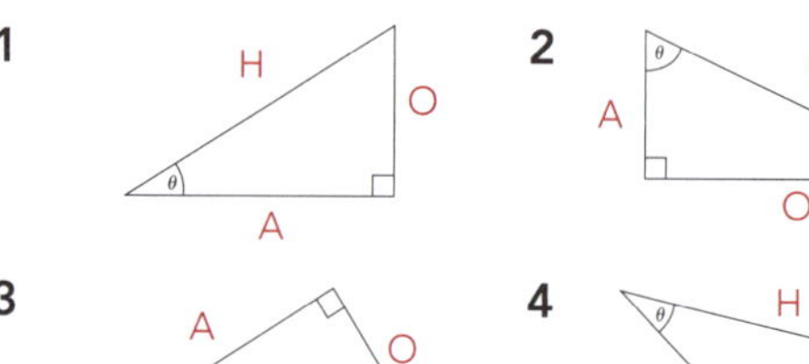

2 (A, H, O)

4 (H, A, O)

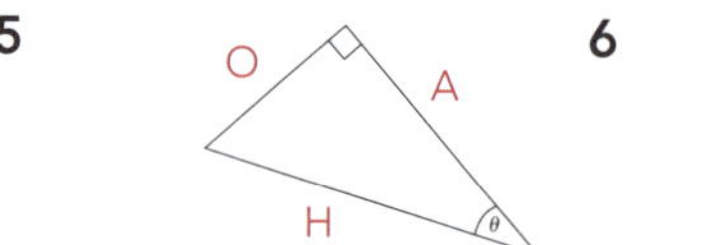

6 (A, H, O)

Finding short sides using sine and cosine (pp. 13–15)

1 11.87 cm
2 7.419 cm
3 32.48 mm
4 77.63 mm
5 3.011 m
6 8.940 km
7 398.1 km
8 321.9 mm
9 0.8954 m
10 40.38 cm

Finding long sides (hypotenuses) using sine and cosine (pp. 16–17)

1 26.42 cm
2 16.51 cm
3 138.2 mm
4 214.4 mm
5 6.055 m
6 30.42 km
7 1039 km

Finding sides using tangents (pp. 18–19)

1 8.748 cm
2 22.40 cm
3 120.9 mm
4 171.2 mm
5 4.052 m
6 28.93 km
7 439.3 km

Finding angles (pp. 20–21)

1 47.5°
2 29.7°
3 52.3°
4 49.1°
5 57.4°
6 65.0°
7 25.5°
8 54.6°
9 35.2°

Putting it together — with Pythagoras (pp. 22–24)

1 31.6°
2 9.793 cm
3 34.44 mm
4 49.1°
5 a 4.8° b 12.04 units
6 a 5.492 cm b 56.5°
7 a 9.970 cm b 7.962 cm
8 a 26.6° b 8.941 cm
9 ∠BAC = 48° (180° – 105°– 27°)
AD = 3.546 cm
DC = 8.019 cm
AC = 11.57 cm
10 ∠BYX = 45° (isos Δ)
$h = 25$ cm
11 11.52 cm
12 48.25 cm

ISBN: 9780170370394

Angles with lines (pp. 25–33)

Revision (p. 25)

Term	Description	Diagram
Acute angle	An angle that is less than 90°	
Obtuse angle	An angle that is greater than 90°, but less than 180°	
Right angle	An angle that is 90°	
Reflex angle	An angle that is greater than 180°, but less than 360°	
Scalene triangle	A triangle with no equal sides	
Isosceles triangle	A triangle with exactly two equal sides	
Equilateral triangle	A triangle with three equal sides	
Right-angled triangle	A triangle in which one angle is 90°	
Supplementary angles	Two angles that add to 180°	
Complementary angles	Two angles that add to 90°	

Fundamentals (pp. 26–27)

1 $a = 55°$ (vert opp ∠s =)
$b = 85°$ (∠s on a line = 180°)
$c = 40°$ (∠s in Δ = 180°)

2 $a = 45°$ (vert opp ∠s =)
$b = 25°$ (∠s on a line = 180°)
$c = 110°$ (∠s in Δ = 180°)
$d = 120°$ (∠s at a point = 360°)

3 $a = 45°$ (∠s on a line = 180°)
$b = 60°$ (∠s at a point = 360°)
$c = 105°$ (ext ∠ of Δ = sum of int opp ∠s)

4 $a = 88°$ (vert opp ∠s =)
$b = 114°$ (ext ∠ of Δ = sum of int opp ∠s)
$c = 114° - 88°$
$= 26°$ (ext ∠ of Δ = sum of int opp ∠s)

5 $a = 40°$ (∠s on a line = 180°)
$b = 50°$ (∠s in Δ = 180°)
$c = 69°$ (vert opp ∠s =)
$d = 61°$ (∠s in Δ = 180°)

Developing a chain of reasoning (pp. 28–29)

1 ∠CED = 93° (∠s on a line = 180°)
∠DCE = 48° (∠s in Δ = 180°)
∴ $\theta = 48°$ (vert opp ∠s =)

2 ∠EBF = 44° (∠s on a line = 180°)
∠BEF = 46° (vert opp ∠s =)
∠EFB = 90° (∠s in Δ = 180°)
∴ $\theta = 67°$ (∠s at a point = 360°)

3 ∠DCG = 75° (vert opp ∠s =)
∠EDG = 114° (ext ∠ of Δ = sum of int opp ∠s)
∴ $\theta = 114° + 31°$
$= 145°$ (ext ∠ of Δ = sum of int opp ∠s)

4 ∠BAD = 40° (∠s on a line = 180°)
∠BCD = 50° (∠s in Δ = 180°)
∴ $\theta = 111° - 50°$
$= 61°$ (ext ∠ of Δ = sum of int opp ∠s)

5 ∠ECD = 43° (vert opp ∠s =)
∠CEF = 123° (ext ∠ of Δ = sum of int opp ∠s)
∠EFG = 85° (∠s at a point = 360°)
∠FGE = 123° – 85°
= 38° (ext ∠ of Δ = sum of int opp ∠s)
∴ $\theta = 38°$ (vert opp ∠s =)

Parallel lines (pp. 30–33)

1 Alternate
2 Co-interior
3 Co-interior
4 Corresponding
5 Corresponding
6 Alternate
7 Co-interior
8 Corresponding

Calculating angles where parallel lines are involved (pp. 31–33)

1 $a = 50°$ (co-int ∠s add to 180°, // lines)
$b = 65°$ (corr ∠s =, // lines)
$c = 65°$ (∠s on a line = 180°)

2 $a = 92°$ (co-int ∠s add to 180°, // lines)
$b = 36°$ (alt ∠s =, // lines)
$c = 128°$ (∠s in Δ = 180° and ∠s on a line = 180°)

3 $a = 109°$ (∠s in Δ = 180°)
$b = 109°$ (corr ∠s =, // lines)
$c = 26°$ (alt ∠s =, // lines)

ISBN: 9780170370394

4 $a = 47°$ (corr ∠s =, // lines)
$b = 99°$ (∠s on a line = 180°)
$c = 34°$ (corr ∠s =, // lines)

5 $a = 48°$ (alt ∠s =, // lines)
$b = 48°$ (alt ∠s =, // lines)
$c = 119°$ (co-int ∠s add to 180°, // lines)
$d = 71°$ (∠s on a line = 180° and ∠s in Δ = 180°)

6 $a = 26°$ (alt ∠s =, // lines)
$b = 135°$ (∠s on a line = 180°)
$c = 19°$ (∠s in Δ = 180°)

7 ∠AFE = 47° (alt ∠s =, // lines)
∠ECF = 64° (∠s on a line = 180°)
∴ $\theta = 111°$ (ext ∠ of Δ = sum of int opp ∠s)

8 ∠BEF = 36° (co-int ∠s add to 180°, // lines)
∴ $\theta = 94°$ (∠s on a line = 180°)

9 ∠BDC = 31° (∠s in Δ = 180°)
∠ADC = 66° (co-int ∠s add to 180°, // lines)
∴ $\theta = 66° - 31° = 35°$

10 ∠FGC = 99° (corr ∠s =, // lines)
∠GDF = 27° (∠s in Δ = 180°)
∴ $\theta = 153°$ (∠s on a line = 180°)

11 ∠FAD = 65° (∠s in Δ = 180°)
∠ECD = 65° (corr ∠s =, // lines)
∴ $\theta = 115°$ (∠s on a line = 180°)

12 ∠FCB = 147° (co-int ∠s add to 180°, // lines)
∠FCD = 116° (co-int ∠s add to 180°, // lines)
∴ $\theta = 97°$ (∠s at a point = 360°)

13 ∠ABC = 117° (co-int ∠s add to 180°, // lines)
∠BCD = 117° (alt ∠s =, // lines)
∠FDC = 63° (co-int ∠s add to 180°, // lines)
∠FDE = 101° (co-int ∠s add to 180°, // lines)
∴ $\theta = 101° - 63° = 38°$

Polygons (pp. 34–42)

Triangles (pp. 34–35)

1 $a = 50°$ (isos Δ, base ∠s =)
$b = 80°$ (∠s in Δ = 180°)
$c = 50°$ (alt ∠s =, // lines)

2 $a = 41°$ (co-int ∠s add to 180°, // lines)
$b = 56°$ (isos Δ, base ∠s =)
$c = 83°$ (∠s on a line = 180°)

3 $a = 66°$ (∠s on a line = 180°)
$b = 57°$ (isos Δ, base ∠s = and ∠s in Δ = 180°)
$c = 57°$ (isos Δ, base ∠s =)
$d = 123°$ (∠s on a line = 180°)

4 $a = 68°$ (alt ∠s =, // lines)
$b = 56°$ (isos Δ, base ∠s = and ∠s in Δ = 180°)
$c = 56°$ (isos Δ, base ∠s =)

5 ∠CBD = 60° (equilat Δ)
∴ $\theta = 81°$ (∠s on a line = 180°)

6 ∠CAB = 44° (isos Δ, base ∠s =)
∠DAC = 80°
∴ $\theta = 50°$ (isos Δ, base ∠s =)

7 ∠BCD = 64° (∠s on a line = 180°)
∠DBC = 52° (isos Δ, base ∠s =)
∴ $\theta = 33°$ (alt ∠s =, // lines)

Quadrilaterals (pp. 36–38)

Name	Angles	Side lengths	Parallel sides	Diagonals
Square	Right angles at each vertex	Four equal	Two pairs, opposite each other	Equal, perpendicular, bisect each other
Rhombus	No right angles, two pairs of equal angles opposite each other	Four equal	Two pairs, opposite each other	Not equal, perpendicular, bisect each other
Rectangle	Four right angles	Two equal pairs, opposite each other	Two pairs, opposite each other	Equal, not perpendicular, bisect each other
Parallel-ogram	No right angles, two pairs of equal angles opposite each other	Opposite sides equal, adjacent sides not	Two pairs, opposite each other	Not equal, not perpendicular, bisect each other
Kite	One pair of equal angles opposite each other, remaining angles not equal	Two pairs of equal sides, adjacent to each other	No parallel sides	Not equal, perpendicular, one is bisected by the other
Trapezium	No right angles, no equal angles	None equal	One pair parallel sides, opposite each other, remaining sides not parallel	Not equal, not perpendicular, do not bisect each other
Isosceles trapezium	Two pairs of equal angles adjacent to each other	One pair of equal sides, opposite each other, remaining sides not equal	One pair parallel sides, opposite each other, remaining sides not parallel	Equal, not perpendicular, do not bisect each other

1 $a = 71°$ (∠s of quad = 360°)
$b = 239°$ (∠s at a point = 360°)
$c = 55°$ (∠s of quad = 360°)

2 $a = 117°$ (alt ∠s =, // lines)
$b = 117°$ (opp ∠s of parallelogram =)
$c = 63°$ (∠s on a line = 180°)

ISBN: 9780170370394

$d = 63°$ (opp ∠s of parallelogram =)

3 $a = 114°$ (symmetry — opp ∠s of kite =)
$b = 27°$ (congruent Δs)
$c = 78°$ (∠s of quad = 360°)

4 $a = 124°$ (∠s on a line = 180°)
$b = 124°$ (base ∠s isos trapezium =)
$c = 56°$ (co-int ∠s add to 180°, // lines)
$d = 56°$(co-int ∠s add to 180°, // lines)

5 $a = 90°$ (diags of rhombus ⊥)
$b = 27°$ (rhombus is bisected by diagonals)
$c = 126°$ (co-int ∠s add to 180°, // lines)

6 ∠ADC = 208° (∠s at a point = 360°)
∠BCD = 31° (symmetry — opp ∠s of kite =)
∴ $\theta = 90°$

7 ∠AED = 56° (∠s on a line = 180°)
EA = ED (diag of rectangle equal and bisected)
∴ $\theta = 62°$ (∠s in Δ = 180° and isos Δ, base ∠s =)

8 ∠ADC = 67° (alt ∠s =, // lines)
∠FCD = 60° (∠s in Δ = 180°)
∴ $\theta = 16°$ (∠s on a line = 180°)

9 Obtuse ∠BED = 127° (∠s of quad = 360°)
Reflex ∠BED = 233° (∠s of quad = 360°)
∴ $\theta = 28°$ (∠s of quad = 360°)

10 ∠EDC = 78° (co-int ∠s add to 180°, // lines)
∠DCB = 102° (co-int ∠s add to 180°, // lines)
∴ $\theta = 22°$ (∠s of quad = 360°)

Exterior and interior angles of polygons (pp. 41–42)

1 112° **2** 67°
3 141° **4** 95°
5 110° **6** 90°
7 265° **8** 75°
9 273° **10** 72°
11 36° **12** 140°

3D shapes (pp. 43–46)

1 **a** HC = 53 cm
b EC = 59.94 cm
c ∠ECH = 27.8°
d ∠CHG = 31.9°
e ∠DHC = 58.1°

2 **a** AD = 5.809 m
b ∠TDA = 43.4°
c AC = 4.212 m
d TC = 6.928 m
e ∠TCA = 52.6°

3 **a** TB = 7.349 m
b ∠TBD = 54.7°
c Angle = 63.4°
d Angle = 180° – 2(65.9°) = 48.2°

Bearings (pp. 47–51)

1 **a** ∠ABC = 90°
b AC = 20 miles
∴ Course is 48 miles
c ∠BCA = 53.1°
d Bearing = 256.9°

2 **a** ∠ABC = 90°
b AC = 9.8 miles
∴ Course is 9.761 miles
c ∠BCA = 38.0°
d Bearing = 214°

3 **a** AC = 244.4 m
∴ Course is 579.4 m
b Bearing = 087.2°

4 **a** QR = 901.4 m
∴ Course is 2151.4 m
b Bearing = 019.3°

5 **a** ∠ABC = 58°
b AC = 969.6 m
∴ Course is 2969.6 m
c ∠BCA = 61°
d Bearing = 234°

6 **a** BC: 107°
CD: 179°
DE: 251°
EA: 323°
b 114.6 m

Angles with circles (pp. 52–65)

Fundmentals (pp. 52–55)

1 $a = 37°$ (isos Δ, base ∠s =)
$b = 106°$ (∠s in Δ = 180°)
$c = 254°$ (∠s at a point = 360°)

2 $a = 27°$ (∠s on the same arc =)
$b = 64°$ (∠s on the same arc =)
$c = 91°$ (ext ∠ of Δ = sum of int opp ∠s)

3 $a = 124°$ (∠ at cent is 2 x ∠ at circ)
$b = 236°$ (∠s at a point = 360°)
$c = 21°$ (∠s of quad = 360°)

4 $a = 51°$ (isos Δ, base ∠s =)
$b = 51°$ (isos Δ, base ∠s =)
$c = 39°$ (∠ in semicircle = 90°)
$d = 39°$ (isos Δ, base ∠s =)

5 $a = 38°$ (isos Δ, base ∠s =)
$b = 104°$ (∠s in Δ = 180°)
$c = 124°$ (∠s at a point = 360°)
$d = 28°$ (isos Δ, base ∠s =)
$e = 24°$ (isos Δ, base ∠s =)

6 $a = 61°$ (∠s on the same arc =)
$b = 51°$ (∠ in semicircle = 90°)
$c = 29°$ (∠s in Δ = 180°)
$d = 51°$(∠s on the same arc =)

7 $\angle$ABC = 90° ($\angle$ in semicircle = 90°)
$\angle$ABD = 36° (isos Δ, base $\angle$s =)
∴ θ = 54°

8 $\angle$ADB = 96° ($\angle$ at cent is 2 x $\angle$ at circ)
reflex $\angle$ADB = 264°
∴ θ = 28° ($\angle$s of quad = 360°)

9 $\angle$BDC = 78° ($\angle$s on a line = 180°)
$\angle$BCD = 51° (isos Δ, base $\angle$s =)
∴ θ = 39° ($\angle$ in semicircle = 90°)

10 $\angle$ACD = 97° ($\angle$s on the same arc =)
∴ θ = 49° ($\angle$s in Δ = 180°)

11 $\angle$ACD = 49° ($\angle$ at cent is 2 x $\angle$ at circ)
$\angle$CED = 61° ($\angle$s in Δ = 180°)
$\angle$BEA = 61° (vert opp $\angle$s =)
∴ θ = 21° ($\angle$s in Δ = 180°)

Angles involving tangents (pp. 56–58)

1 a = 33° (tan ⊥ rad)
b = 33° (isos Δ, base $\angle$s =)
c = 114° ($\angle$s in Δ = 180°)

2 a = 60° ($\angle$ between chord and tan = $\angle$ in the alt seg)
b = 101° ($\angle$s in Δ = 180°)

3 a = 52° ($\angle$ between chord and tan = $\angle$ in the alt seg)
b = 76° (isos Δ, base $\angle$s =, = tans)
c = 78° ($\angle$s in Δ = 180°)

4 a = 53° ($\angle$ between chord and tan = $\angle$ in the alt seg)
b = 106° ($\angle$ at cent is 2 x $\angle$ at circ)
c = 37° (isos Δ, base $\angle$s =)
d = 37°(isos Δ, base $\angle$s =)
e = 16° ($\angle$s in quad Δ = 360°)

5 a = 65° ($\angle$ between chord and tan = $\angle$ in the alt seg)
b = 65° (alt $\angle$s =, // lines)
c = 50° ($\angle$s in Δ = 180°)

6 $\angle$CBE = $\angle$CDE = 90° (tan ⊥ rad)
$\angle$BED = 116° ($\angle$s of quad = 360°)
∴ θ = 58° ($\angle$ at cent is 2 x $\angle$ at circ)

7 $\angle$BAC = 65° ($\angle$ between chord and tan = $\angle$ in the alt seg)
obtuse $\angle$AFB = 116° ($\angle$ at cent is 2 x $\angle$ at circ)
$\angle$BAF = 32° (isos Δ, base $\angle$s =)
∴ θ = 33°

8 $\angle$DAB = 59° ($\angle$ between chord and tan = $\angle$ in the alt seg)
$\angle$ABD = 80° ($\angle$s in Δ = 180°)
∴ θ = 80° ($\angle$s on the same arc =)

9 $\angle$EAD = $\angle$EDA = 46° (isos Δ, base $\angle$s =)
$\angle$ADB = 44° (tan ⊥ rad)
$\angle$ABD = 92° ($\angle$s in Δ = 180° and isos Δ, base $\angle$s =)
∴ θ = 46° ($\angle$ at cent is 2 x $\angle$ at circ)

Angles in cyclic quadrilaterals (pp. 59–61)

1 a = 93° (ext $\angle$ of cyclic quad = the int opp $\angle$)
b = 91° (opp $\angle$s of cyclic quad add to 180°)

2 a = 75° ($\angle$ at cent is 2 x $\angle$ at circ)
b = 75° (ext $\angle$ of cyclic quad = the int opp $\angle$)
c = 210° ($\angle$s at a point = 360°)
d = 37° ($\angle$s of quad = 360°)

3 a = 90° ($\angle$ in semicircle = 90°)
b = 90° ($\angle$ in semicircle = 90°)
c = 45° (isos Δ, base $\angle$s =)
d = 37° (ext $\angle$ of cyclic quad = the int opp $\angle$)
e = 53° ($\angle$s in Δ = 180°)

4 a = 64° (isos Δ, base $\angle$s =)
b = 64° (isos Δ, base $\angle$s =)
c = 116° (opp $\angle$s of cyclic quad add to 180°)
d = 71° (ext $\angle$ of cyclic quad = the int opp $\angle$)

5 a = 60° (opp $\angle$s of cyclic quad add to 180°)
b = 30° (isos Δ, base $\angle$s =)
c = 90° ($\angle$ in semicircle = 90°)
d = 120° (ext $\angle$ of cyclic quad = the int opp $\angle$)

6 $\angle$BAE = 57° (ext $\angle$ of cyclic quad = the int opp $\angle$)
$\angle$AFE = 90° ($\angle$ in semicircle = 90°)
$\angle$EAF = 25° ($\angle$s in Δ = 180°)
∴ θ = 82°

7 $\angle$BCD = 54° (opp $\angle$s of cyclic quad add to 180°)
$\angle$ECD = 33° (isos Δ, base $\angle$s =)
$\angle$ECB = 21° (isos Δ, base $\angle$s =)
∴ θ = 21°

8 $\angle$DAB = 76° ($\angle$ at cent is 2 x $\angle$ at circ)
$\angle$EDC = 50° (co-int $\angle$s add to 180°, // lines,
$\angle$DCB = 104° (opp $\angle$s of cyclic quad add to 180°)
∴ θ = 54° ($\angle$s of quad = 360°)

9 $\angle$ACD = 90° ($\angle$ in semicircle = 90°)
$\angle$CDA = 50° ($\angle$s in Δ = 180°)
$\angle$BDA = 24° (opp $\angle$s of cyclic quad add to 180°)
∴ θ = 24°

10 $\angle$AED = 117° ($\angle$ at cent is 2 x $\angle$ at circ)
$\angle$AFD = 126° ($\angle$s at a point = 360°)
$\angle$ABD = 63° (opp $\angle$s of cyclic quad add to 180°)
$\angle$FDE = 58.5° ($\angle$s of quad = 360° and opp $\angle$s of kite =)
∴ θ = 84.5°

ISBN: 9780170370394

Concyclic points (pp. 62–65)

1 ∠BAD + ∠DCB = 179°
∴ A, B, C and D are not concyclic because opposite angles of a cyclic quadrilateral must add to 180°.

2 ∠DAC ≠ ∠DBC
∴ A, B, C and D are not concyclic because angles on the same arc (CD) must be equal.

3 ∠AED = 49° (∠s on a line = 180°)
∠ADE = 103° (∠s in Δ = 180°)
∴ A, B, C and D are concyclic because angles on the same arc (AB) are both 103°.

4 A, B, C and D are not concyclic because for a quadrilateral to be cyclic, the exterior angle (71°) must equal the interior opposite angle, which is 109°, not 71°.

5 ∠BAE = 42° (alt ∠s =, // lines)
∠ABE = 44° (∠s in Δ = 180°)
∴ A, B, C and D are not concyclic because angles on the same arc (AD) must be equal, but ∠BAE = 42° and ∠ABE = 44°.

6 ∠DCB = 104° (isos Δ, base ∠s = and ∠s in Δ = 180°)
∠DAB = 76°
∴ ∠DCB + ∠DAB = 180°
∴ A, B, C and D are concyclic because opposite angles of a cyclic quadrilateral must add to 180°.

7 A, B, D and E are concyclic because ∠ABD = ∠AED = 90°. These add to 180°, so A, B, D and E must be the corners of a cyclic quadrilateral.

8 ∠FEC = 68° (∠s in Δ = 180°)
∴ C, D, E and F are concyclic because angles on the same arc (CF) are both 68°.

9 ∠EAB = 76°
∠EDB = 104°
∴ ∠EAB + ∠EDB = 180°
∴ E, A, B and D are concyclic because opposite angles of a cyclic quadrilateral must add to 180°.

10 ∠BCA = 24° (∠s in Δ = 180°)
∴ ∠BCD = 112°
∠BED = 68°
∴ ∠BCD + ∠BED = 180°
∴ B, C, D and E are concyclic because opposite angles of a cyclic quadrilateral must add to 180°.

11 ∠BFA = 65° (vert opp ∠s =)
∠ABF = 43° (∠s in Δ = 180°)
∴ A, B, C and E are concyclic because angles on the same arc (AE) are both 43°.

Use of algebra in geometry (pp. 66–68)

1 $a = x$ (corr ∠s =, // lines)
$c = x$ (isos Δ, base ∠s =)
∴ $b = 180° - 2x$ (∠s in Δ = 180°)

2 $c = 2x$ (∠ at cent is 2 x ∠ at circ)
$a = b$ (isos Δ, base ∠s =)
∴ $b = \frac{180° - 2x}{2} = 90° - x$ (∠s in Δ = 180°)

3 ∠ABD = x (∠s on the same arc =)
∠BAC = x (alt ∠s =, // lines)
∴ ∠AED = $2x$ (ext ∠ of Δ = sum of int opp ∠s)

4 ∠EBC = ∠EDC = 90° (tan ⊥ rad)
∠BED = $360° - 2(90°) - x$
$= 180° - x$ (∠s of quad = 360°)
∴ ∠BAD = $\frac{180° - x}{2}$ (∠ at cent is 2 x ∠ at circ)

5 $a = x$ (isos Δ, base ∠s =)
$b = 90°$ (∠ in semicircle = 90°)
$c = 90° - x$ (∠s in Δ = 180°)
$d = 180° - (90° - x)$
$= 90° + x$ (∠s on a line = 180°)
∴ $e = 180° - (90° + x) - x$
$= 90° - 2x$ (∠s in Δ = 180°)

6 $a = 180° - x$ (∠s on a line = 180°)
$b = \frac{360° - (180 - x) - y}{2}$
$= \frac{180° + x - y}{2}$ (∠s of quad = 360°)

7 $b = x$ (∠ between chord and tan = ∠ in the alt seg)
$c = 90° - x$ (tan ⊥ rad)
∴ $a = 180° - y - x - (90° - x)$ (∠s in Δ = 180°)
$= 90° - y$

8 $e = 180° - x - y$ (∠s on a line = 180°)
$c = 180° - x - y$ (isos Δ, base ∠s =)
$b = 180° - y$ (co-int ∠s add to 180°, // lines)
∴ $a = 180° - (180° - x - y) - (180° - y)$
$= 2y + x - 180°$ (∠s on a line = 180°)

9 ∠BCA = $\frac{180° - x}{2}$ (∠ sum isos Δ = 180°)
∠ADC = $180° - x$ (opp ∠s of cyclic quad add to 180°)

∠ACD = $180° - y - (180° - x)$
$= x - y$ (∠s in Δ = 180°)
∴ ∠BCD = $(x - y) + \frac{180° - x}{2}$
$= 90° - y + \frac{x}{2}$

ISBN: 9780170370394

10 $c = 180° - x$ (co-int ∠s add to 180°, // lines)
$b = 180° - y - (180° - x)$ (∠s on a line = 180°)
$= x - y$
∴ $a = 180° - (x - y)$ (co-int ∠s add to 180°, // lines)
$= 180° - x + y$

Similar triangles (pp. 69–73)

1 ∠ABC = ∠DFE = 55°
∴ ΔABC is similar to ΔDFE

2 ∠BAC = 60°
∴ ΔABC is similar to ΔFED

3 ∠ABC = 86°, ∠DEF = 87°
∴ ΔABC is not similar to ΔFED

4 ∠ABC = 26°
∴ ΔABC is similar to ΔDFE

5 ∠EAC = ∠EDC (∠s on the same arc =)
∠B is common to both triangles
∴ ∠ACB = ∠DEB (∠s in Δ = 180°)
∴ ΔABC is similar to ΔDBE

6 $\frac{10}{5} = \frac{8}{4} = \frac{6}{3} = 2$
∴ ΔABC is similar to ΔTSR

7 $\frac{50}{40} = \frac{35}{28} = \frac{35}{28} = 1.25$
∴ ΔABC is similar to ΔTRS

8 $\frac{600}{400} = \frac{381}{254} = 1.5 \quad \frac{570}{395} = 1.44$
∴ ΔABC is not similar to ΔTRS

9 ∠BAC = 60° (∠s in Δ = 180°)
∴ ΔBAC is similar to ΔEFD because the angles are equal.
$\frac{x}{6.9} = \frac{7.2}{6.3} \Rightarrow x = 7.886$ cm

10 ΔBAC is similar to ΔEFD because the angles are equal.
$\frac{x}{10.7} = \frac{6.8}{13.5} \Rightarrow x = 5.390$ cm

11 ΔBAC is similar to ΔDFE because the angles are equal.
$\frac{x}{4.9} = \frac{3.5}{4.5} \Rightarrow x = 3.8\dot{1}$ cm

12 ∠C is common to both triangles.
∠CBD = ∠CAE (corr ∠s =, // lines)
∴ Δ CBD is similar to Δ CAE because the angles are equal.
$\frac{25}{35} = \frac{45}{45 + x} \Rightarrow x = 18$ cm

13 ∠ABE = ∠ADE (∠s on the same arc =)
∠BAD = ∠DEB (∠s on the same arc =)
∴ ΔBAC is similar to ΔDEC because the angles are equal.
$\frac{x}{3} = \frac{8}{6} \Rightarrow x = 4$ cm

14 ∠B is common to both triangles.
∠BAC = ∠BED (∠s on the same arc =)
∴ ΔBAC is similar to ΔBED because the angles are equal.
$\frac{6.3}{6.9 + x} = \frac{6.9}{5.4 + 6.3} \Rightarrow x = 3.783$ cm

15 ∠E is common to both triangles.
∠EFJ = ∠EGK (corr ∠s =, // lines)
∴ ΔEFI is similar to ΔEGH because the angles are equal.
$\frac{4}{9} = \frac{3.5}{3.5 + x} \Rightarrow x = 4.375$ cm

16 ∠DCB = ∠AFE = 90° (given)
∠CBD = ∠CAE (corr ∠s =, // lines)
∠CAE = ∠AEF (alt ∠s =, // lines)
∴ ΔCBD is similar to ΔFEA because the angles are equal.
$\frac{3.5}{6} = \frac{x}{4} \Rightarrow x = 2.333$ cm

17 ∠ECD = ∠BCA (vert opp ∠s =)
∠DEC = ∠CBA (alt ∠s =, // lines)
∴ ΔDEC is similar to ΔABC because the angles are equal
$\frac{x}{6} = \frac{4}{5} \Rightarrow x = 4.8$ cm

Proof (pp. 74–77)

1 **Step 1:** Construct a line parallel to AB and ED, which passes through C. Call this GH.
Step 2: Use parallel line rules.

∠ABC = ∠BCH (alt ∠s =, // lines)
∠EDC = ∠DCH (alt ∠s =, // lines)

∠BCD = ∠BCH + ∠DCH
∴ **∠BCD = ∠ABC + ∠EDC**

2 **Step 1:** Show that triangles CAD and CEB are similar.

∠C is common to both triangles
∠CAD = ∠CEB (∠s on the same arc =)
∴ ∠CDA = ∠CBE (∠s in Δ = 180°)
∴ ΔCAD is similar to ΔCEB

Step 2: Because the triangles are similar, their sides are in proportion.
ΔCAD is similar to ΔCEB so:
$$\frac{CA}{CE} = \frac{CD}{CB}$$
∴ **CA x CB = CE x CD**

ISBN: 9780170370394

3

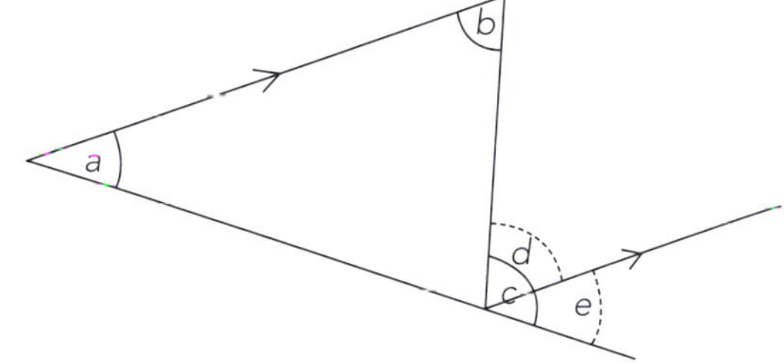

$\angle d = \angle b$ (alt $\angle$s =, // lines)
$\angle e = \angle a$ (corr $\angle$s =, // lines)
$\angle c = \angle d + \angle e$
$\therefore$ **$\angle c = \angle a + \angle b$**

4 **Step 1:** Show that triangles BAC and BDA are similar.

$\angle$B is common to both triangles.
$\angle$BAC = $\angle$BDA ($\angle$ between chord and tan = $\angle$ in the alt seg)
$\therefore$ $\angle$BCA = $\angle$BAD ($\angle$s in Δ = 180°)
$\therefore$ ΔBAC is similar to ΔBDA

Step 2: Because the triangles are similar, their sides are in proportion.

ΔBAC is similar to ΔBDA so:

$$\frac{BA}{BD} = \frac{BC}{BA}$$

$\therefore$ **BA^2 = BC x BD**

5 $\angle a = \angle b$ (isos Δ, base $\angle$s =)
$\angle c = \angle d$ (isos Δ, base $\angle$s =)
$\angle e = \angle f$ (isos Δ, base $\angle$s =)
From the large triangle:
$2\angle b + 2\angle c + 2\angle e = 180°$ ($\angle$s in Δ = 180°)
So $\angle b + \angle c + \angle e = 90°$
$\therefore$ **$\angle b + \angle c = 90° - e$**
From the bottom triangle:
$\angle i = 180° - 2\angle e$
$= 2(90° - \angle e)$
$\therefore$ **$\angle i = 2(\angle b + \angle c)$**

6 $a = b$ (isos Δ, base $\angle$s =)
$c = d$ (isos Δ, base $\angle$s =)
$\therefore$ $e = 2c$ (ext $\angle$ of Δ = sum of int opp $\angle$s)
Similarly, $f = 2b$.
However, $e + f = 180°$.
$\therefore$ $2c + 2b = 180°$
$\therefore$ **$c + b$** $= 90°$

Practice questions (pp. 78–83)

1a i $\angle$BAE = $\angle$BEA = 64° (isos Δ, base $\angle$s = and $\angle$s in Δ = 180°)
$\therefore$ $\angle x = 64°$ (corr $\angle$s =, // lines)

ii A trapezium has two parallel sides; and the remaining two sides are not parallel.
$\angle x = 64°$ and $\angle y = 116°$ — these are co-interior angles that add to 180°, so AD is parallel to BC.
AB and DC are clearly not parallel.
$\therefore$ William is correct; ABCD is a trapezium.

iii $a = b = \frac{180° - f}{2}$ (isos Δ, base $\angle$s = and $\angle$s in Δ = 180°)

$g = 180° - \frac{180° - f}{2}$ ($\angle$s on a line = 180°)
$= 90° + \frac{f}{2}$

b i $\angle$FDE = $\angle$FED (ext $\angle$ of regular pentagon = 72°)
$\therefore$ $\angle y = 36°$ ($\angle$s in Δ = 180°)

ii $\angle$DEA = 108° (int $\angle$ of regular pentagon)
$\therefore$ $\angle$GDE = $\angle$GED = 54° ($\angle$s in Δ = 180°)
$\therefore$ $\angle z = 72°$ ($\angle$s in Δ = 180°)

iii $\angle$AGD = 144° ($\angle z = 72°$ above)
$\angle y = 36°$ (see **i**)
But 144° + 36° = 180° $\Rightarrow$ AGDF is a cyclic quadrilateral because these would be opposite angles.
(*Or* $\angle$GAE = 54°
$\angle$GDF = 54° + 72° = 126°
But 54° + 126° = 180° $\Rightarrow$ AGDF is a cyclic quadrilateral because these would be opposite angles.)

2 a i Course = 4981 m

ii $\tan^{-1}(\frac{1200}{1700}) = 35.2°$
Due south is 180°, bearings go clockwise, so bearing from C to A must be 180° + 35.2° = 215.2°

b i Course is 2160 m
ii Bearing is 267.1°
iii Bearing is 308.5°

c $\angle$ACB = 38°
$\angle$CAB = 52°
BC = 793.6 m
AC = 1007 m
$\therefore$ Course = 2421 m

3 a i Angle EDC is common to ΔCDE and ΔADB
$\angle$ABC = $\angle$CED (given)
$\therefore$ $\angle x = \angle y$ ($\angle$s in Δ = 180°)

ii CD = 7.5 cm
iii AE = 6.5 cm

b i $x = 244°$ (ext $\angle$ of cyclic quad = the int opp $\angle$ and $\angle$s at a point = 360°)
ii $y = 68°$ ($\angle$s on a line = 180° and $\angle$s of quad = 360°)

ISBN: 9780170370394

iii ∠BAD = s (ext ∠ of cyclic quad = the int opp ∠)
∠BFD = 360° – 2s (∠ at cent is 2 x ∠ at circ and ∠s at a point = 360°)
By symmetry, ∠t = ∠ABF (BF = DF (radii) and AB = AD (given))

$\therefore \angle t = \frac{360° - (360° - 2s) - s}{2} = \frac{s}{2}$

c **i** Yes, she is correct.
∠DBC = 90° (tan ⊥ rad)
From ΔDFB: 180° = ∠DBF + 90° + 55° (∠s in Δ = 180°)
From line ABC: 180° = ∠DBF + 90° + x (∠s on a line = 180°)
∴ **x = 55°**

ii ∠DBF = ∠p (see reasoning in **i**)
∠DBF = ∠EDF (DF is a perpendicular bisector of isos Δ)
∴ ∠EDB = 2 x ∠BDF
= 2p
But ∠EDB = 2 x ∠EGB (∠ at cent is 2 x ∠ at circ)
∴ 2p = 2q
∴ **p = q**

ISBN: 9780170370394